Surface/Volume

Alan E. Rubin

Surface/Volume

How Geometry Explains Why Grain
Elevators Explode, Hummingbirds Hover,
and Asteroids are Colder than Ice

 Springer

Alan E. Rubin
Department of Earth, Planetary, and Space
Sciences
University of California
Los Angeles, CA, USA

ISBN 978-3-031-23748-5   ISBN 978-3-031-23749-2 (eBook)
https://doi.org/10.1007/978-3-031-23749-2

This Springer imprint is published by the registered company Springer Nature Switzerland AG
The registered company address is: Gewerbestrasse 11, 6330 Cham, Switzerland

A nanosphere is very small
It is a very tiny ball
Its atoms—a surficial field
But little volume is concealed.

The planet Earth is quite immense
The heat inside is quite intense
It takes so very long to cool
The S/V ratio makes this rule.

For Geometers
and those who move in their circles

Introduction

Hardy tourists can hike up the Kīlauea volcano on the big island of Hawai'i. It is an 8.5-km round trip from the Visitor Center along Crater Rim Trail and around the Kīlauea Iki Trail loop. The terrain is rugged with deep cracks and razor-sharp lava. Kīlauea is probably the most active volcano in the world, erupting, on average, every two to three years. A series of magma chambers beneath Kīlauea (at depths of 3 km, 30 km, and 90 km) provides the molten rock. During eruptions, low-viscosity basaltic lava pours from volcanic vents and flows downslope. At night, visitors can see a river of thin, slowly undulating red-orange stripes surrounding dark amoeboid slabs of congealed lava.

Whether Kīlauea is erupting or not, once a month, if skies are clear, the full moon bathes the volcano in pale white light. The Moon's nearside is covered with basaltic lava plains similar in composition to basalts from Kīlauea. But the mantle of the Moon is cold; major volcanic activity on the Moon ended about 2½ billion years ago.

Why do volcanoes erupt every day on Earth, but not on the Moon?

It is a matter of size and geometry. The radius of the Earth is 3.7 × that of the Moon. The surface area of the Earth (continents and oceans combined) is about 13 × that of the Moon; the volume of the Earth is 49 × that of the Moon. The difference in volume far exceeds the difference in surface area (49 vs. 13). This is because volume increases as the third power of radius (r × r × r) while surface area increases only as the second power (r × r). This geometric constraint ensures that the (surface-area)/volume ratio of the Earth is less than 30% that of the Moon. The Earth therefore has a lot of volume (in which heat can be generated), but comparatively little surface area from which interior heat can escape. Temperatures in the Earth's lithosphere can exceed 1300°C; hence, the active volcanoes. The Moon has a much smaller volume and a relatively high amount of surface area; this geometric property allowed most of the Moon's internal heat to escape into space early in Solar System history.

The surface-area/volume ratio (commonly simplified as surface/volume) constrains much of the physical world. Small objects lose heat faster than large ones. This general property applies not only to planets and moons but also to mammals and birds. Small animals lose heat quickly. To avoid shivering or freezing to death, small vertebrates must maintain a high metabolic rate. While an elephant's heart may beat 25 times a minute, a shrew's heart beats more than 600 times a minute. To go about its daily activities, a house mouse (two-to-three times the size of a shrew) requires about 40 times as much energy per unit mass as an elephant.

Surface/volume ratios also influence the structures of plants. Leaves from deciduous trees are broad and flat with high surface/volume ratios. The large surface area captures a lot of sunlight; the thin leaf structure assures that the Sun's rays do not have far to travel before entering the chloroplasts to facilitate photosynthesis.

The rates of chemical processes are also governed by geometry. Example 1—Dissolution: Small soluble substances (e.g., salt crystals) have high surface/volume ratios and dissolve readily. A large fraction of the molecules in the crystals is near the surface and can be easily dislodged when soluble crystals are immersed in water. Example 2—Evaporation: The amount of liquid that evaporates from an open container is a direct function of the surface area of the exposed liquid. As an illustration, let's put two open-top containers atop the dry scrub in California's Mojave Desert; each container holds 1 cubic meter of water. The first container is a cube, 1 m on a side; the exposed surface is 1 m^2. The second container is shaped like a giant sheet pan: 10 m × 10 m × 1 cm; the exposed surface is 100 m^2. Water from the second container evaporates 100 times faster than from the cube (even though both containers initially held the same amount of water).

Nanoparticles have extremely high surface/volume ratios, these tiny objects are nearly all surface with little enclosed volume. A particle 1 nm (a billionth of a meter) across has a surface/volume ratio eight million times greater than that of a pea. Because surface forces predominate in nanometer-size metal particles, these particles melt at much lower temperatures than centimeter-size metal nuggets.

Because mass, like volume, is a three-dimensional property, many physical structures are constrained by their surface/mass ratios. Architect Frank Lloyd Wright's dream of a mile-high skyscraper in Chicago will never materialize. It would be too difficult for such a structure to support its own weight, even though the building was designed to narrow toward the top.

The effective strength of muscles is also constrained by the surface/mass ratio. Raw muscle strength is directly related to a muscle's cross-sectional area, increasing as the square of length. Muscle mass, however, increases as the cube of length. The result is that large birds have such heavy muscles they lack the power to hover, but diminutive hummingbirds have no trouble at all.

Scientists can deduce basic principles from the geometric constraints imposed on physical structures: Small bodies lose heat faster than large bodies. Chemical reactions proceed more rapidly in small objects. Large bodies tend to be so heavy that steps must be taken to reduce the weight—this accounts for the high proportions of (low-density) spongy bone in dinosaurs and the multiple-mirror designs of many modern giant optical telescopes.

The concept of the surface/volume ratio is of enormous importance in understanding the physical world. It explains how rapidly a warm object will lose heat to its surroundings—whether the object is a planet or a gerbil. It governs how efficiently sunlight can penetrate leaves and facilitate photosynthesis. It accounts for the numerous folds in the human brain, explains why our small intestine is all coiled up and why krill have finely branched appendages. It shows why we are much more likely to get frostbite on our toes and fingers than our legs and arms. And it warns us to be cautious around grain elevators.

Other Books by Alan E. Rubin

Disturbing the Solar System: Impacts, Close Encounters, and Coming Attractions. Princeton University Press, 2002, Princeton, New Jersey. 376 pp.

Son of Man: A Personal Memoir by Yehoshuah ben Yahweh, Translated and Edited by Arthur Melton and Monica Wheatley. A novel. 2015, CreateSpace. 147 pp.

With Chi Ma: *Meteorite Mineralogy.* Cambridge University Press, 2021, Cambridge. 404 pp.

Contents

1 A Truncated History of Geometry 1
    References 10

2 Middle-School Math 11
    2.1 Geometric Forms 11
    2.2 Plane and Solid Geometry 12
         2.2.1 Squares and Cubes 12
         2.2.2 Circles and Spheres 15
    2.3 The Ratio of Surface Area to Volume 19
    Reference 23

3 Asteroids, Moons, Planets, and Meteorites 25
    3.1 Inventory of the Solar System 25
    3.2 Surface/Volume Effects in the Inner Solar System 28
    3.3 Meteorite Strewn Fields 33
    3.4 Chondrules 36
    3.5 Aqueous Alteration of Chondritic Meteorites 40
    3.6 Cosmic Spherules 42
    References 43

4 Geologic Processes 45
    4.1 Extrusive and Intrusive Igneous Rocks 45
    4.2 Permeability of Sandstones 50
    4.3 Erosion and Weathering 52
    References 57

5 Geometry of Life 59
    5.1 Metabolism 59
    5.2 The Surface/Volume Ratios of Mammals and Birds 62
    5.3 Muscles and Wings 71
    5.4 Formidable Formicidae or the Mighty Ant 75
    5.5 Biological Rules of Thumb 77

		5.5.1	Bergmann's Rule	77
		5.5.2	Allen's Rule	79
		5.5.3	Hesse's Rule	79
	5.6	Gigantothermy		80
	5.7	The Impossibility of King Kong (But Not Dinosaurs)		81
	5.8	Thermoregulation of Insects		82
	5.9	The Risk of Animal Dehydration		84
	5.10	Krill		87
	5.11	Sponges		88
	5.12	Leaves and Trees		91
		5.12.1	Leaves	91
		5.12.2	Deciduous Trees	94
	5.13	Human Anatomy		97
		5.13.1	The Brain	97
		5.13.2	The Respiratory Tract	98
		5.13.3	The Gastrointestinal Tract	102

5.14 Human Anatomical and Behavioral Responses to Extreme
Temperatures ... 105

5.15 Frostbite ... 106
5.16 Bacteria ... 108
5.17 Oxygen Diffusion Through Red Blood Cells ... 112
References ... 114

6 Biochemistry ... 117
6.1 Protein Folding ... 117
6.2 A Summing Up ... 122
References ... 123

7 Chemical Reactions ... 125
7.1 Elementary Chemistry ... 125
7.2 Chemical Properties of Water ... 129
7.3 Dissolution ... 131
7.4 The Effort to Reduce Sodium in Food ... 133
7.5 Sugar and Candy Making ... 135
7.6 Evaporation ... 136
7.7 Osmosis ... 138
7.8 Grain Elevators ... 141
References ... 144

8 Ecology ... 147
8.1 Wildfires ... 147
8.2 Coral Reefs ... 151
References ... 153

9 **Manufacturing** 155
    9.1   Artificial Bones 155
    9.2   Artificial Lungs 157
    9.3   Aerogel and *Stardust* 158
    9.4   Giant Telescope Mirrors 161
    9.5   Nanoparticles 165
    References 171

Epilogue 173

Index 177

Chapter 1
A Truncated History of Geometry

Administrating a government costs money—taxes must be assessed. By the beginning of the Third Millennium BCE, Egyptians had developed methods for measuring the surface area of land holdings so proprietors could pay their fair share. Length was measured by the cubit (the length of the arm from the elbow to the tip of the middle finger), palm (width) and hand (width including the thumb). Eventually, length was standardized as a cubit rod—essentially a yardstick. Surveyors measured land dimensions using ropes tied with knots at fixed intervals and stretching the ropes until taut. Let's say a rectangular plot owned by Mr. Sneferukhaf is 300 cubits wide and 400 cubits long. What is the area? Area was measured in units of setats, eventually set at one setat equals one square khet, where 1 khet = 100 cubits. Sneferukhaf's plot of land is 3 × 4 khets in size, equivalent to 12 setats (or 120,000 square cubits).

Sneferukhaf's neighbor, Mr. Khaemwaset, has a circular field 400 cubits across. Egyptians estimated the area of a circle by that of a square with sides equal to eight-ninths of the circle's diameter. This effectively yielded a value for π of 3.16 (deviating from the actual value by only 0.6%). The diameter of Khaemwaset's field is 4 khets; the side of the prescribed square is (8/9) × 4 khets and the area of the square (and thus of the circular field) is about 12 ¾ setats. The actual area is 125,664 square cubits.

The Egyptians called this type of applied mathematics "earth measurement"—*geometry* in Greek.

Somewhat later, perhaps a thousand years before Pythagoras was born, the Babylonians discovered what is essentially the Pythagorean Theorem. As described by Mlodinow (2001), a clay tablet excavated in Assyria read: "Four is the length and five is the diagonal. What is the breadth? Its size is not known. Four times four is sixteen. Five times five is twenty-five. You take sixteen from twenty-five and there remains nine. What times what shall I take in order to get nine? Three times three is nine. Three is the breadth."

A. E. Rubin, *Surface/Volume*, https://doi.org/10.1007/978-3-031-23749-2_1

Fig. 1.1 The 3-4-5 triangle

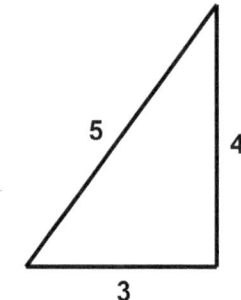

To put it more succinctly:

$$5^2 - 4^2 = b^2$$
$$25 - 16 = b^2$$
$$9 = b^2$$
$$3 = b \text{ (where b is breadth)}$$

This describes a special right triangle (Fig. 1.1), the only one with sides corresponding to consecutive integers.

The Greek merchant and Pre-Socratic philosopher, Thales of Miletus (c. 624–c. 548 BCE), traveled to Babylon and spent time in Egypt, soaking up geometric knowledge. He recognized that, in a plane, those polygons that could be rotated and superimposed on one another are equal (that is, congruent)—they have the same shape irrespective of their movements. Thales understood the properties of similar triangles—these figures have different sizes but have the same shape and same interior angles. The (occasionally) reliable Third Century CE biographer, Diogenes Laërtius, related that Thales measured the height (formally, altitude) of the great pyramids using similar triangles. At the particular moment (in the morning or afternoon) when Thales' shadow was equal in length to his height (perhaps as measured by a Thales-size walking stick), Thales measured the length of the shadow of the pyramid across the ground, concluding that this value (plus half the length of the pyramid's base) equaled the pyramid's height (Fig. 1.2).

Thales also determined that a circle is bisected by its diameter (because the diameter is a straight line going through the center of the circle), that the interior angles of an isosceles triangle are equal (Fig. 1.3a) and that, when any two straight lines intersect, they form two pairs of equal angles (Fig. 1.3b).

He developed what has come to be known as Thales' Theorem: If A, B and C are distinct points on a circle and line AC is the diameter, then angle ABC is a right angle (i.e., 90°) (Fig. 1.4).

Toward the end of his life, Thales met his most brilliant pupil – Pythagoras of Samos (c. 570–c. 495 BCE)—and told the young man to go to Egypt. Pythagoras obliged, but found Egyptian mathematics too practical, too down-to-Earth. He

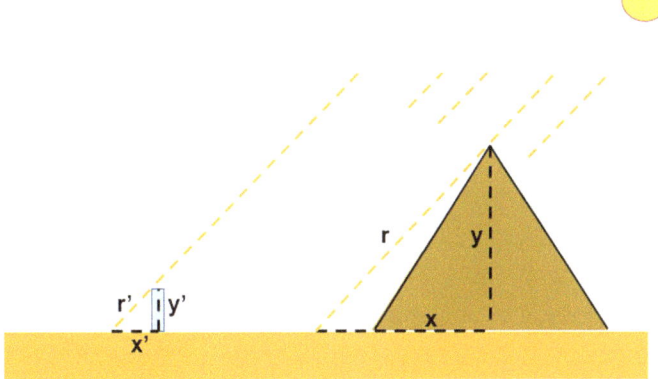

Fig. 1.2 Measuring the height of a pyramid. At the time of day when the length of the shadow of a post is equal to the length of the post itself, the length of the shadow of a pyramid is equal to its height (after adding half the length of the pyramid's base). The right triangles created by the two shadows are similar; hence, the ratios of their sides are the same: $x'/y' = x/y$

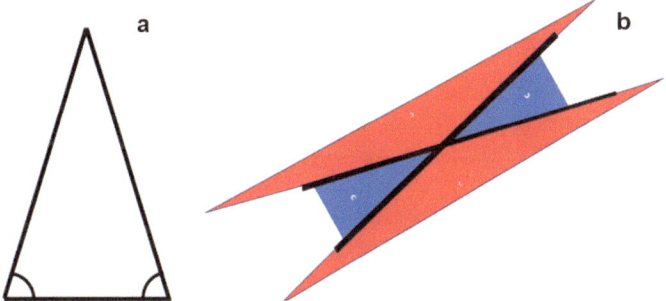

Fig. 1.3 **a** An isosceles triangle. The two bottom angles are equal; the two long sides are equal. **b** Two intersecting straight lines form two pairs of equal angles. The two acute angles (blue) are equal; the two obtuse angles (red) are equal. The blue and red angles are supplementary in that one red angle plus one blue angle sum to 180°

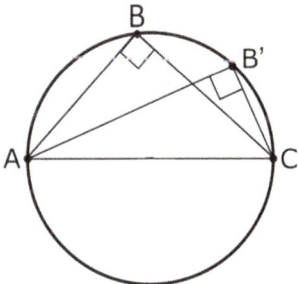

Fig. 1.4 An illustration of Thales' Theorem. If line AC is the diameter of a circle and B and B′ are any points on the circle, distinct from points A and C, then angle ABC (and AB′C) are right angles

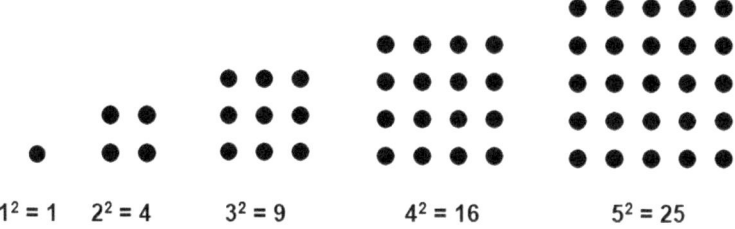

$1^2 = 1$ $2^2 = 4$ $3^2 = 9$ $4^2 = 16$ $5^2 = 25$

Fig. 1.5 Squares formed by pebbles (or dots) as identified by the Pythagoreans

yearned for a higher level of abstraction, one from which results could be derived that were applicable to numerous situations. Around 530 BCE, Pythagoras set up a school/commune (called the Semicircle) in Kroton in southern Italy (a city colonized by Greeks). His group, known as the Pythagoreans, are credited (justly or not) with several mathematical discoveries including, of course, the Pythagorean Theorem, which can be stated as *the sum of the squares of the sides of a right triangle equals the square of the hypotenuse.*

The Fifth Century CE philosophical writer, Proclus Lycius, credited Pythagoras as recognizing the five regular polyhedrons (also known as Platonic solids[1]): the tetrahedron (four faces), cube (six faces), octahedron (eight faces), dodecahedron (twelve faces), and icosahedron (twenty faces). These are the only geometric figures in which all the faces are congruent (having the same size and shape), all the angles are congruent, all the edges are congruent, and the same number of faces meet at every vertex.

The Pythagoreans arranged pebbles in equal numbers of rows and columns to form squares, discovering that only certain numbers were up to the task: 1, 4, 9. 16, 25, 36, etc., (Fig. 1.5). That is why these integers are known as "squares" or "perfect squares".

Pythagoras found that each perfect square is equal to the sum of preceding odd numbers: $1 = 1$ (the first odd number, or 1^2); $4 = 1 + 3$ (the sum of the first two odd numbers or 2^2); $9 = 1 + 3 + 5$ (sum of the first three odd numbers or 3^2); $16 = 1 + 3 + 5 + 7$ (sum of the first four odd numbers or 4^2); $25 = 1 + 3 + 5 + 7 + 9$ (sum of the first five odd numbers or 5^2)....

Although the Babylonians had calculated the length of the diagonal of a square to six decimal places, the Pythagoreans realized the exact value could not be represented by an integer or by the ratio of two integers (hence the label "irrational" number). This shocking concept was incommensurate with Pythagoras's philosophy and his followers were sworn to secrecy. [The modern value of the diagonal of a square of unit length is the square root of 2, symbolized by $\sqrt{2}$, and approximated by 1.414 (or to 50 decimal places, 1.41421356237309504880168872420969807856967187537694...).]

---

[1] In his dialog, *Timaeus,* Plato associated these five regular solids with the classical elements of earth (cube), air (octahedron), water (icosahedron) and fire (tetrahedron). Plato suggested the dodecahedron was used by the gods to arrange the constellations in the heavens.

Fig. 1.6 Diagram showing Eratosthenes' method for measuring the size of the Earth. When two parallel lines are cut by a transversal, alternate interior angles are equal

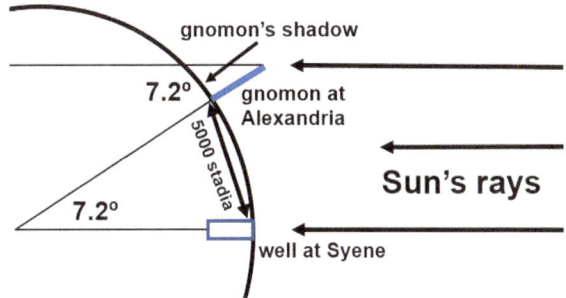

Pythagoras deduced the Earth is a sphere. The only two permanent objects in the sky with a discernable angular diameter are the Sun and Moon, and both are round. By analogy, Earth should be round as well. The sphere was widely viewed in the Greek world as the perfect shape—all points on the surface are equidistant from the center.

Aristotle (384–322 BCE) also accepted the sphericity of Earth, but unlike Pythagoras, provided arguments to support this contention: (1) When masted ships sail over the horizon, they disappear hull first, masts still visible. If the Earth were flat, the entire ships would just decrease in apparent size as they sailed away. (2) During lunar eclipses (when the Earth is directly between the Sun and Moon), the Earth casts a curved shadow across the Moon's surface. (3) Different constellations appear at different latitudes; while traveling north, for example, an observer would find some constellations in the southern sky dipping below the horizon, while more northerly constellations appeared higher in the sky (i.e., they moved toward the south).

Eratosthenes (c. 276–c. 195 BCE), chief librarian of the library of Alexandria, was able to measure the size of the Earth using geometry. On the summer solstice in c. 230 BCE at local noon, he observed the Sun close to the zenith, i.e., directly over a deep well in Syene, Egypt (now Aswan). In Alexandria at noon at the next summer solstice, a vertical rod (a gnomon) cast a shadow. Eratosthenes measured the length of the shadow, and using geometry, found that the Sun was 7.2° south of the zenith (Fig. 1.6). He hired professional surveyors who determined the distance between the two Egyptian cities to be 5000 stadia. Eratosthenes assumed the cities were approximately on a north–south line, that the Sun is so far away that its rays are essentially parallel, and that the Earth is a sphere. He divided 7.2° by 360° and found the distance between these two cities was 2% of the Earth's circumference (7.2/360 = 0.02). He multiplied the linear distance between the cities by 50 (because 50 × 2% = 100%) and calculated the circumference of the Earth to be 250,000 stadia. There were several different stadia in use at the time, but many modern scholars believe the one used by Eratosthenes was equivalent to 184.8 m. This gives a circumference of the Earth of 46,200 km, only about 15% higher than the actual equatorial circumference of 40,075 km.

The greatest of the Greek geometers was Euclid (c. 325–c. 265 BCE). He also lived in Alexandria, collecting, cataloging, and systematizing Greek knowledge about

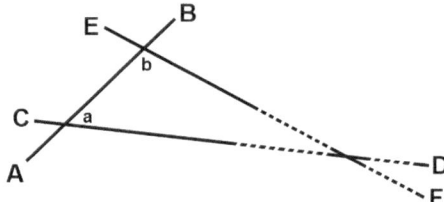

Fig. 1.7 An illustration of Euclid's fifth postulate. If angles a and b sum to less than 180°, then the two lines (C–D and E–F) forming these angles with the first line (A–B) eventually meet on the same side as these angles

geometry. In his greatest work, *Elements*, a book consisting of 13 scrolls, Euclid tried to create a discipline free of unsupported assumptions. He wrote down 23 definitions, five postulates, and five "common notions"; from these he was able to construct proofs of 465 theorems.

His occasionally wordy definitions include those of a point (that which has no part), line (breadthless length), straight line (a line which lies evenly with the points on itself), surface (that which has length and breadth only), obtuse angle (an angle greater than a right angle), acute angle (angle less than a right angle), circle (plane figure contained by one line such that all the straight lines falling upon it from one point among those lying within the figure equal one another), square (quadrilateral figure that is both equilateral and right angled), equilateral triangle (trilateral figure with three sides equal), and parallel lines (straight lines, which, being in the same plane and being produced indefinitely in both directions, do not meet one another in either direction).

The five postulates are: (1) A straight line segment can be drawn between any two points. (2) Straight line segments can be extended indefinitely into straight lines. (3) A circle can be described with any center and radius. (4) All right angles are equal. (5) If two straight line segments intersect a third line segment, forming two interior angles on the same side so that their sum is less than two right angles, then the two straight line segments, if extended indefinitely, will meet on the side on which the angles sum to less than two right angles (e.g., Fig. 1.7).

Euclid's five common notions are: (1) Things which equal the same thing also equal one another. [If A = C and B = C, then A = B.] (2) If equals are added to equals, then the wholes are equal. [If A = B, then A + C = B + C.] (3) If equals are subtracted from equals, then the remainders are equal. [If A = B, then A − D = B − D.] (4) Things which coincide with one another equal one another. (5) The whole is greater than the part.

Euclid's *Elements* is the most influential textbook ever written. More than a thousand editions have been published, second only to that of the Bible (not a textbook). *Elements* has been an integral part of the math curriculum in the West for centuries. The English philosopher John Locke (1632–1704) considered Euclid's *Elements* a template for deciding which moral statements concerning right and wrong rested upon foundations as secure as those of mathematics. Abraham Lincoln (1809–1865)

Fig. 1.8 Relationship between a sphere and a circumscribed right circular cylinder according to Archimedes. Modified from a diagram by Andertxuman

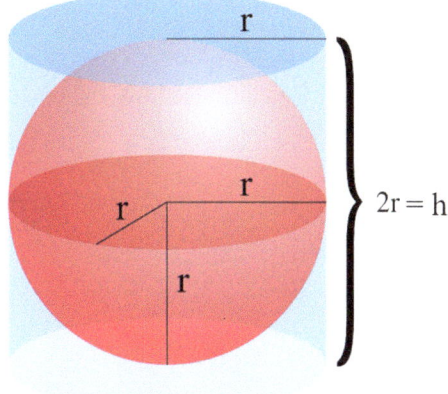

found a copy of *Elements* in his father's house and became well acquainted with it. The mathematical logic in this book significantly influenced aspects of Lincoln's "personal, political, legal, literary, and oratorical accomplishments" (Emerson, 2011). Another admirer was Albert Einstein (1879–1955); although his general theory of relativity was framed on non-Euclidean geometry, Einstein wrote that Euclid's assertions could "be proved with such certainty that any doubt appeared to be out of the question. The lucidity and certainty made an indescribable impression on me."

In c. 225 BCE, Archimedes of Syracuse (c. 287 – c. 212 BCE), the great Greek polymath, derived the formulas for the surface area ($4\pi r^2$) and volume ($4/3\ \pi r^3$) of a sphere by comparing spheres and circumscribed cylinders. He found that the surface area and volume of a sphere are two-thirds that of its circumscribed right circular cylinder (including the bases of the cylinder) (Fig. 1.8). [The surface area and volume of a right circular cylinder are: $SA = 2\pi rh + 2\pi r^2$; $V = \pi r^2 h$, where h is the height of the cylinder. In the case of an inscribed sphere, $h = 2r$.] Archimedes was so proud of this derivation, he requested a diagram resembling Fig. 1.8 be inscribed on his tomb. The request was granted after Archimedes was killed by a Roman soldier during the invasion of Syracuse.

After the fall of the Roman Empire, many of the intellectual achievements of Greece were lost and forgotten in Western Europe, although some works by Aristotle were still to be found in Church libraries. Fortunately, Islamic scholars had copied and preserved ancient Greek writings. They were reintroduced into Europe in the Eleventh to Thirteenth Centuries CE during the Crusades. These works helped pave the way for the Enlightenment, a period some scholars date as beginning in 1637 with the publication of *Discourse on the Method* by René Descartes (1596–1650). It was in this work that Descartes declared: *Je pense, donc je suis* (I think, therefore I am). One of the "rules of thought" followed by Descartes is "never to accept anything for true which I did not know to be such; that is to say, carefully to avoid precipitancy and prejudice, and to comprise nothing more in my judgment than what was presented to my mind so clearly and distinctly as to exclude all ground of doubt."

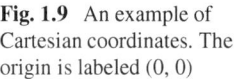

Fig. 1.9 An example of
Cartesian coordinates. The
origin is labeled (0, 0)

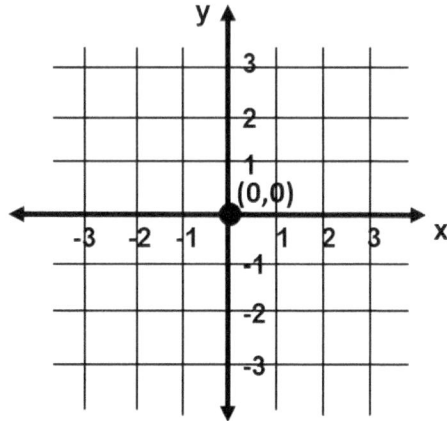

Descartes transformed plane geometry into a graph. He drew a vertical line (called
the y axis) and a horizontal line (called the x axis) meeting at a point called the
origin. Each point on the graph was assigned a y value and an x value, written as an
"ordered pair" (x, y); the origin is (0, 0). Points below and to the left of the origin were
assigned negative values, for example ($-$ 3, $-$ 2). Such graphs are now commonly
referred to as Cartesian coordinates (Fig. 1.9), in homage to Descartes. This branch
of mathematics, in which plane geometry is turned into a map, is called analytic
geometry.

Descartes combined geometry with algebra, enabling him to simplify some of
Euclid's wordy definitions. He described a circle as: all x and y satisfying $x^2 + y^2 =$
r^2 for some constant number r. Alternatively, for a circle with center (a, b) and radius
r, the equation is $(x - a)^2 + (y - b)^2 = r^2$.

The points on any straight line on the graph are related to each other. A straight
line can be defined by the expression: $ax + by + c = 0$ where a, b and c are real
numbers (including integers, fractions, and irrational numbers) and a and b are not
both zero. Any values of a, b and c that satisfy the equation (as long as a and b are
not both zero) lie on the same line; any values that do not satisfy the equation will
not be on the same line.

Our truncated history of geometry ends with Galileo Galilei (1564–1642), the
first thinker to recognize the importance of the surface/volume ratio. The discussion
of the topic appears in Galileo's final book, *Two New Sciences*, covering the strength
of materials (i.e., material engineering) and the motion of objects (i.e., kinematics).
The book is written as a four-day dialog among three men (Simplicio, Sagredo,
and Salviati), eponymous personifications of simplicity, sacredness, and salvation,
representing the early, middle, and latter stages of Galileo's evolving understanding.

In the discussion on Day 2, Salviati explains that as the length of a beam increases,
it becomes less able to support a load plus its own weight (Fig. 1.10). This is because,
as the beam gets larger, the increase in its volume and mass are much greater than
its increase in surface area. While surface area increases as the second power of

Fig. 1.10 Figure 17 from
Galileo's *Two New Sciences.*
Prism ABCD, attached to a
vertical wall at point AB, is
supporting a weight E at the
other end. If the beam gets
too long, it will not be able to
support weight E and its own
weight

length, volume and mass increase as the third power. This is the "square-cube law".
In Proposition I, Salviati states "A prism or solid cylinder of glass, steel, wood or
other breakable material which is capable of sustaining a very heavy weight when
applied longitudinally is, as previously remarked, easily broken by the transverse
application of a weight which may be much smaller in proportion as the length of the
cylinder exceeds its thickness." Whereas a short beam could support the weight E in
Fig. 1.10 as the beam grows longer (maintaining its same thickness and density), at
some point, the beam will fail and fracture.

   The same principle applies to pieces of chalk. The standard blackboard chalk is
a cylinder 80 mm long and 9 mm wide, made mainly from calcium carbonate or
gypsum. It fractures easily. A glance at the shelf beneath classroom blackboards
(at least those blackboards that remain in the modern era) typically reveals several
broken pieces of chalk; the only intact ones have not yet been used. The stress put on
the 80-mm-long chalk cylinder by grasping it firmly and writing commonly breaks
it in two. The smaller chalk pieces are quite stable, however, and can withstand
appreciable stress. The intact pieces of chalk behave like Galileo's over-extended
beam.

   Galileo also used his Salviati character to analyze the strength of bones. "...nor
can nature produce trees of extraordinary size because the branches would break
down under their own weight; so also it would be impossible to build up the bony
structures of men, horses, or other animals so as to hold together and perform their
normal functions if these animals were to be increased enormously in height; for

this increase in height can be accomplished only by employing a material which is harder and stronger than usual, or by enlarging the size of the bones, thus changing their shape until the form and appearance of the animals suggest a monstrosity."

Salviati continued. "Clearly then, if one wishes to maintain in a great giant the same proportion of limb as that found in an ordinary man, he must either find a harder and stronger material for making the bones, or he must admit a diminution of strength in comparison with men of medium stature; for if his height be increased inordinately, he will fall and be crushed under his own weight. Whereas, if the size of a body be diminished, the strength of that body is not diminished in the same proportion; indeed the smaller the body, the greater its relative strength. Thus, a small dog could probably carry on his back two or three dogs of his own size; but I believe that a horse could not carry even one of his own size."

After the Roman Inquisition tried Galileo in 1633 and found him "vehemently suspect of heresy", he was enjoined from publishing any new works, even those unrelated to the heliocentric theory (wherein the Earth revolves around the Sun). Galileo remained under house arrest until his death in 1642. *Two New Sciences* was published in Leiden in the Netherlands in 1638. This was a predominantly Protestant country, less susceptible to the influence of the Inquisition. Despite Galileo's legal predicament, the book was allowed to be sold in Rome's bookstores. It sold out quickly.

In 1979, more than three centuries after Galileo's death, Pope John Paul II suggested that "theologians, scholars and historians, animated by a spirit of sincere collaboration, will study the Galileo case more deeply and in loyal recognition of wrongs, from whatever side they come."

References

Emerson, J. (2011). Abraham Lincoln and the structure of reason. *Civil War Book Review, 13*(1). https://doi.org/10.31390/cwbr.13.1.09

Mlodinow, L. (2001). *Euclid's window: the story of geometry from parallel lines to hyperspace.* Free Press, 306 pp.

Chapter 2
Middle-School Math

2.1 Geometric Forms

There are only a few common geometric forms, each with distinct properties and practical uses:

Bricks are **rectangular prisms**, block-shaped objects good for stacking.

Triangular prisms made of glass can refract white light, separating it into its constituent rainbow colors—red, orange, yellow, green, blue, indigo, and violet (ROYGBIV).

Billiard balls are **spheres** with no sharp corners; they transfer their kinetic energy to other balls upon impact.

Pyramids are great for interring dead pharaohs, easing their transition to the afterlife.

Cones are ideal for holding scoops of ice cream or sherbet.

Tortillas are thin, flat **disks** (short cylinders), perfect for wrapping enchiladas without significantly affecting the flavor of the filling.

The fluted Doric columns holding up the ceiling of the Parthenon are large **cylinders**; these pillars transmit their heavy load, by axial compression, to the structural elements beneath them.

The formulas for calculating the geometric properties of these forms are independent of size.

For example, the surface area of any sphere is $4 \times \pi \times r^2$ (where r is the radius) regardless of whether r is 1 mm (as in small pearls), 50 m (as in the Spaceship Earth sphere at Disney's EPCOT theme park in Florida) or 100 lightyears (as in large globular clusters distributed around galaxies).

The volume of any rectangular prism is $l \times w \times h$ (length $\times$ width $\times$ height); the formula pertains to blocks where the sides are all the same (e.g., sugar cubes and dice) as well as those where the dimensions are very different: A floor joist commonly used in home construction in the United States has dimensions of ~ 2 $\times$ 4 $\times$ 96 inches (~ 5 $\times$ 10 $\times$ 244 cm); one of the tallest skyscrapers in Chicago is

© The Author(s), under exclusive license to Springer Nature Switzerland AG 2023
A. E. Rubin, *Surface/Volume*, https://doi.org/10.1007/978-3-031-23749-2_2

the Aon Center (formerly the Standard Oil Building) with a height that is nearly six times greater than its length or width.

The fundamental feature explored in this book is the ratio of surface area to volume; it is this special ratio that influences much of the physical world. It is often called the "Surface to Volume Ratio", written simply as "surface/volume" and abbreviated S/V. More properly, it is the "Surface Area to Volume Ratio", commonly abbreviated SA/V. Although, formally, this ratio is expressed as $(unit)^{-1}$ (e.g., μm^{-1}; cm^{-1}; m^{-1}; km^{-1}), in practice, only the numerical value is required to understand the physical relationships. The volume of an object increases as the cube of length (i.e., l^3); the surface increases more slowly, only as the square of length (i.e., l^2). Thus, the larger an object is, the greater the difference between the numerical values of surface area and volume. In short, the (Surface Area)/Volume ratio is much higher for small objects than large ones. This leads to a key geometric observation:

Small objects have lots of surface area and relatively little volume. Large objects have lots of volume and relatively little surface area.

2.2 Plane and Solid Geometry

2.2.1 Squares and Cubes

A square (Fig. 2.1) is a two-dimensional figure, confined (by definition) to a plane. Its consecutive sides are of equal length and there is a right angle (i.e., an angle of 90°) between them. The *area* (**A**) of a square is the product of the length of two sides:

$$Area\ of\ a\ Square = side \times side$$

If each side of a square has a length of 1 cm, the area of the square will be 1 cm $\times$ 1 cm = 1 cm^2. If each side has a length of 2 cm, the area will be 2 cm $\times$ 2 cm = 4 cm^2. One can construct a graph (Fig. 2.2) showing how the area increases as the square of the length:

$$Area = (length)^2$$

A cube (Fig. 2.3) is a three-dimensional solid object bounded by six square faces; three faces meet at each vertex. The cube can be considered a projection of a square into the third dimension out to a depth equivalent to the length of one side.

Cubes abound in nature. Of the seven possible ways in which crystalline blocks can be stacked periodically in three dimensions without producing gaps, one such way is the cube (known in crystallography as the isometric system). Common minerals with this morphology include galena (lead sulfide; PbS) (Fig. 2.4a), pyrite (commonly known as "fool's gold"; FeS_2) (Fig. 2.4b), and halite (table salt; NaCl) (Fig. 2.4c). Isometric minerals with an internal cubic structure can exhibit external forms

Fig. 2.1 "Fifty Squares"

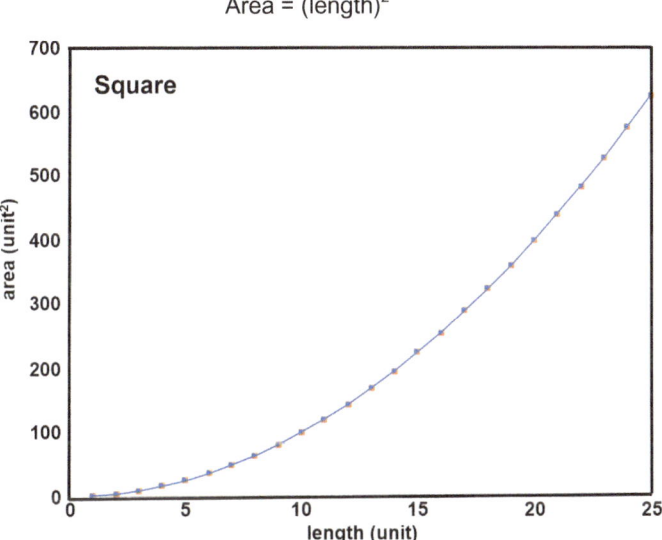

Fig. 2.2 The area of a square grows exponentially with increasing length of a side

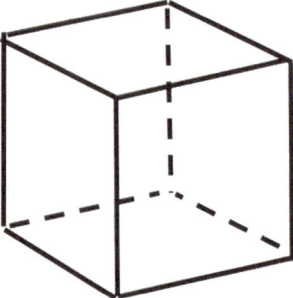

Fig. 2.3 Schematic diagram of a cube, a three-dimensional object depicted in two dimensions using perspective drawing

(called habits) that are not, in fact, cubes. Magnetite (Fe_3O_4), for example, can occur as cubes, octahedrons, and dodecahedrons; in some carbonaceous-chondrite meteorites, small grains of magnetite occur as spherulites, framboids, plaquettes and platelets.

To calculate the area of a cube, we can start by assigning each face (or square side) of the cube a number as a label: Face 1, Face 2, Face 3, Face 4, Face 5, Face 6. Because all faces are equivalent, it doesn't matter where we begin. The total area of any solid object is called the *surface area* (abbreviated **SA**). In the case of a cube, the surface area is the sum of the areas of its six square faces:

$$\text{Surface Area of a Cube} = \text{Area}_{\text{Face 1}} + \text{Area}_{\text{Face 2}} + \text{Area}_{\text{Face 3}} + \text{Area}_{\text{Face 4}} + \text{Area}_{\text{Face 5}} + \text{Area}_{\text{Face 6}}$$

Because the six faces are equivalent, the Surface Area is 6 times the area of any face:

Fig. 2.4 Natural cubic minerals. **a** Galena (lead sulfide; PbS). **b** Pyrite (iron sulfide, "fool's gold"; FeS_2). **c** Halite (sodium chloride, common table salt; NaCl). *Images courtesy* of the Smithsonian Institution

$$\text{Surface Area of a Cube} = 6 \times \text{Area}_{\text{Face}}$$

If each side of the cube has a length of 1 cm, the surface area of the cube is 6 × 1 cm^2 = 6 cm^2. If each side has a length of 2 cm, SA = 6 × 4 cm^2 = 24 cm^2.

Because cubes are three dimensional objects, they take up space. The amount of space enclosed within the cube is the volume (**V**). Volume is equal to the product of the lengths of the height, width, and depth; because these three lengths are equivalent for a cube, V = (side)^3. This can be expressed as:

$$\text{Volume of a Cube} = s^3 \text{ or } l^3$$

(where s stands for side and l stands for length, both signifying the same thing). If each side of a cube has a length of 1 cm, the volume of the cube will be V = (1 cm) × (1 cm) × (1 cm) = 1 cm^3. If each side is 2 cm long, V = (2 cm)^3 or 8 cm^3. If each side is 3 cm, V = (3 cm)^3 or 27 cm^3.

The volume increases as the cube of the length, while the surface area increases only as the square of the length. For any length greater than 6 units, the numerical value of the volume exceeds that of the surface area. For l = 12 units, the numerical value of the volume is twice that of the surface area (Table 2.1). And it doesn't matter if the units are centimeters or inches, kilometers or miles, angstroms, or light years.

2.2.2 *Circles and Spheres*

Now, let's go back to the tabletop, back to two dimensions, and examine the circle. This is a two-dimensional figure in which all points along its boundary in a particular plane are equidistant from a given point. The boundary (or perimeter) of the circle is the *circumference* (**C**). The *diameter* (**d**) of the circle is the length of a straight-line segment (called a chord) that passes from one point on the circumference through the center of the circle and on to the opposite side. The *radius* (**r**) of the circle is half the diameter; it is defined as the length of the segment that connects the center of the circle to any point on its circumference.

The Greek letter *pi*, usually written in the lower case as π, is the symbol for the ratio of a circle's circumference to its diameter:

$$\pi = C/d \text{ or } \pi = C/2r$$

π is the 16th letter of the Greek alphabet and the first letter of the Greek word *perimetros* (circumference). In decimal notation, π is equivalent to approximately 3.14159265358979323846264338327950288841971.... Although π can be approximated by fractions (e.g., 22/7 or 355/113), it is an *irrational number* whose decimals do not terminate or end in a repeating sequence. They continue endlessly. An approximation of π as 3 appears in the Bible (1 Kings 7:23, KJV):

Table 2.1 Geometric
properties of cubes

Cubes

Length	Surface area	Volume	(Surface area)/volume
1	6	1	6
2	24	8	3
3	54	27	2
4	96	64	1.5
5	150	125	1.2
6	216	216	1
7	294	343	0.85714
8	384	512	0.75
9	486	729	0.66667
10	600	1000	0.6
11	726	1331	0.54545
12	864	1728	0.5
13	1014	2197	0.46154
14	1176	2744	0.42857
15	1350	3375	0.4
16	1536	4096	0.375
17	1734	4913	0.35294
18	1944	5832	0.33333
19	2166	6859	0.31579
20	2400	8000	0.3
21	2646	9261	0.28571
22	2904	10,648	0.27273
23	3174	12,167	0.26087
24	3456	13,824	0.25
25	3750	15,625	0.24
50	15,000	125,000	0.12
100	60,000	1,000,000	0.06
500	1,500,000	125,000,000	0.012
1000	6,000,000	1,000,000,000	0.006

If the length is expressed in units, the surface area is unit^2, the
volume is unit^3, and the surface/volume ratio is unit^{-1}

And he made a molten sea, ten cubits from the one brim
to the other (i.e., diameter): it was round all about, and
his height was five cubits: and a line (i.e., circumference)
of thirty cubits did compass it round about.

Rational numbers can be written precisely as the ratio of two integers; irrational numbers cannot. When expressed in decimal form, rational numbers can terminate (e.g., $2/5 = 0.4$; $3/8 = 0.375$; $1/32 = 0.03125$) or continue endlessly as a single repeating digit (e.g., $1/3 = 0.33333333...$; $5/6 = 0.83333333...$; $8/9 = 0.88888888...$) or a series of repeating digits (e.g., $1/82 = 0.0121951219512195...$; $9/44 = 0.204545454545...$; $1,234,567/9999999 = 0.123456712345671234567123456...$). The number of repeating digits can be enormous: for example, $1/61$ has a sequence of 60 decimal digits that repeats endlessly. Here is a single set of the repeating digits:

$1/61 = 0.01639344262295081967213114754098360655737704918032786885 2459...$

When converted to decimal notation, the fraction $1/9857$ has a sequence of 9856 digits (mercifully not listed below) that repeats endlessly.

In contrast, the digits of π continue endlessly with *no discernable pattern* in an apparently statistically random manner. In 2021, a supercomputer in Switzerland calculated π to 62.8 trillion digits.

The value of π is independent of the size of the circle: if one circle has a diameter 25 times larger than a second circle, the first circle's circumference will also be 25 times larger than the second one and the C/d ratio will remain constant. The formula for circumference is:

$$C = 2\pi r \text{ or } C = \pi d$$

The area of a circle is the product of π and the square of the radius:

$$\text{Area of a Circle} = \pi r^2$$

If the radius of a particular circle is 1 cm long, the area of the circle will be: $A = \pi \times (1 \text{ cm})^2 = 3.14159 \text{ cm}^2$. If the radius is 2 cm, the area $A = \pi \times (2 \text{ cm})^2 = 12.56636 \text{ cm}^2$. Neither of these areas is exact because π itself cannot be written as a finite decimal.

A sphere (Fig. 2.5) is a three-dimensional structure wherein every point on its boundary is the same distance from a given point in space. A sphere can be considered the solid produced when a circle is rotated along its diameter through three dimensions. A sphere can be thought of as a perfectly smooth ball.

Spheres or spheroids (triaxial ellipsoids that are nearly spherical) are abundant in nature. Stars and planets are spheroidal because gravity acts as though it's located in the center of such bodies, pulling all the overlying mass toward the center. The Earth has a solid ball of metallic iron-nickel (2442 km in diameter) as its inner core. Liquid droplets are spheroidal (especially in the absence of gravity) due to surface tension; this is the property of liquids allowing them to resist an external force. Because a sphere is the shape with the lowest surface area for a given enclosed volume and because surface tension tends to minimize surface area, droplets tend to form spheres.

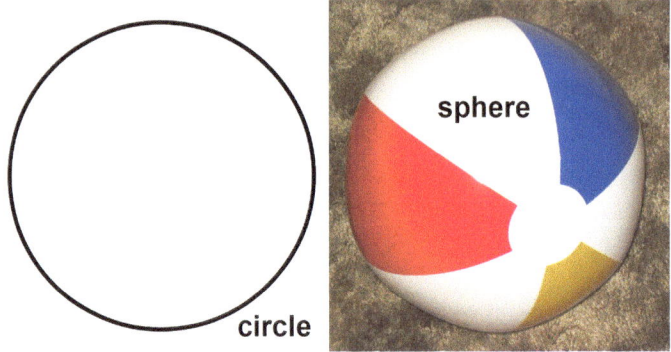

circle

sphere

Fig. 2.5 A circle (an abstract, two-dimensional entity) and a sphere (a real-world, three-dimensional beachball)

The surface area of a sphere is defined by:

$$\text{Surface Area of a Sphere} = 4\pi r^2$$

The radius and diameter of a sphere are analogous to those of the circle: the radius is the distance from a point on the surface of the sphere to the center; the diameter is twice the radius, equivalent to a straight-line segment from a point on the surface of the sphere through the center of the sphere and on to the surface on the opposite side.

If the radius of a particular sphere is 1 cm long, the surface area of the sphere will be: $SA = 4\pi \times (1 \text{ cm})^2 = 12.56636 \text{ cm}^2$. If the radius is 2 cm, $SA = 4\pi \times (2 \text{ cm})^2 = 50.26544 \text{ cm}^2$.

The volume of a sphere is determined by the formula:

$$\text{Volume of a Sphere} = (4/3)\pi r^3$$

If the radius of the sphere is 1 cm, the volume will be: $V = (4/3) \times \pi \times (1 \text{ cm})^3 = 4.1887867 \text{ cm}^3$. If the radius is 2 cm, the volume $V = (4/3) \times \pi \times (2 \text{ cm})^3 = 33.510293 \text{ cm}^3$. If the radius is 3 cm, $V = (4/3) \times \pi \times (3 \text{ cm})^3 = 113.09724 \text{ cm}^3$.

Relative to the radius, the volume of a sphere increases much faster than the surface area (Fig. 2.6), i.e., the volume increases as the cube of the radius while the surface area increases only as the square. For any radius greater than 3 units, the numerical value of the volume exceeds that of the surface area. For r = 6 units, the numerical value of the volume is twice that of the surface area (Table 2.2).

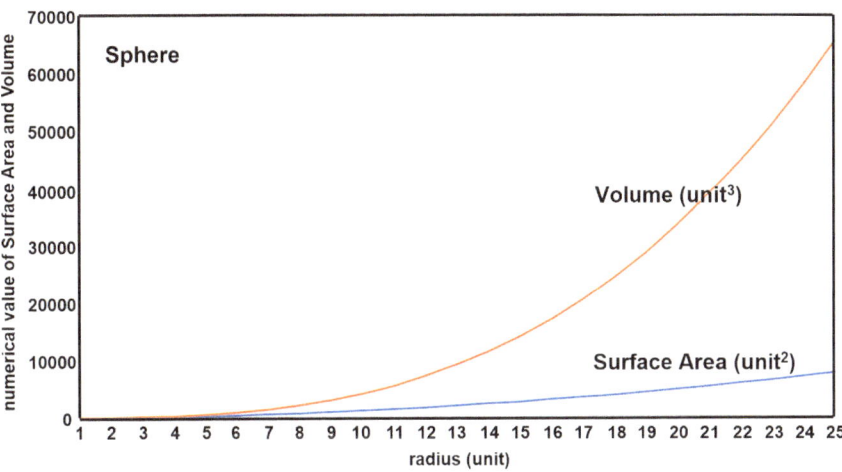

Fig. 2.6 Because surface area increases as the square of the radius while volume increases as the cube of the radius, the (surface area)/volume ratio decreases rapidly with increasing size

2.3   The Ratio of Surface Area to Volume

Because the volume of a set of objects increases much faster than the surface area (Fig. 2.6), the (Surface Area)/Volume ratio is much higher for small objects than for large ones. **Small objects have lots of surface area and little volume. Large objects have lots of volume and relatively little surface area.** This is illustrated in Fig. 2.7, the most important figure in this book.

We can check the numerical values of this ratio for cubes (Table 2.1) and spheres (Table 2.2). If we equate the length of a side of a cube and the radius of a sphere, we see that the numerical value of the surface/volume ratio of a cube is exactly twice that of the sphere. In fact, a sphere has the smallest SA/V ratio of all objects of a given volume; objects with sharp angles, such as four-sided pyramids (tetrahedra), cubes, octahedra, dodecahedra, and icosahedra, have higher SA/V ratios than a sphere for any given volume.

A cube that is 10 cm on a side has a surface/volume ratio that is 10 times smaller than a cube 1 cm on a side. A cube 100 cm on a side has a surface/volume ratio that is 100 times smaller than a cube that is 1 cm on a side. The situation is exactly analogous for spheres: A sphere that has a radius of 10 cm has a surface/volume ratio that is 10 times smaller than a sphere with a radius of 1 cm. A sphere with a radius of 100 cm has a surface/volume ratio that is 100 times smaller than a sphere of radius 1 cm. The same relationships hold for any linear units, be they micrometers or centimeters, kilometers or inches, rods or fathoms, miles or furlongs, yards or versts.

Consequences of the surface/volume ratio permeate our daily lives. A cook may wish to prepare classic garlic bread or parmesan-and-garlic potato wedges; the recipes call for large cloves of garlic. To release the flavor of garlic, the cloves must be minced. A large wedge-shaped clove of garlic (Fig. 2.8) has an average surface area of 0.22

Table 2.2 Geometric properties of spheres

Spheres				
Radius	Diameter	Surface area	Volume	(Surface area)/volume
1	2	12.566	4.189	3
2	4	50.265	33.510	1.5
3	6	113.097	113.097	1
4	8	201.062	268.083	0.75
5	10	314.159	523.599	0.6
6	12	452.389	904.779	0.5
7	14	615.752	1436.755	0.4286
8	16	804.248	2144.661	0.375
9	18	1017.876	3053.628	0.33333
10	20	1256.637	4188.790	0.3
11	22	1520.531	5575.280	0.27273
12	24	1809.557	7238.229	0.25
13	26	2123.717	9202.772	0.23077
14	28	2463.009	11,494.040	0.21429
15	30	2827.433	14,137.167	0.2
16	32	3216.991	17,157.285	0.1875
17	34	3631.681	20,579.526	0.17647
18	36	4071.504	24,429.024	0.16667
19	38	4536.460	28,730.912	0.15789
20	40	5026.548	33,510.322	0.15
21	42	5541.769	38,792.386	0.14286
22	44	6082.123	44,602.238	0.13636
23	46	6647.610	50,965.010	0.13043
24	48	7238.229	57,905.836	0.125
25	50	7853.982	65,449.847	0.12
50	100	31,415.927	523,598.776	0.06
100	200	125,663.706	4,188,790.205	0.03
500	1000	3,141,592.654	523,598,775.6	0.006
1000	2000	12,566,370.614	4,188,790,205	0.003

If the radius and diameter are expressed in units, the surface area is unit^2, the volume is unit^3, and the surface/volume ratio is unit^{-1}

cm^2, a volume of 2.3 cm^3, and a surface/volume ratio of ~ 0.10 cm^{-1} (e.g., El-Gayar & Bahnas, 2005). In contrast, a typical piece of minced garlic is roughly spherical in shape with a diameter of 0.15 cm. Its surface area is 0.071 cm^2, its volume is 0.0018 cm^3, and its surface/volume ratio is ~ 40 cm^{-1}. There would be ~ 1280 pieces of minced garlic produced from a single large clove. The total surface area of the

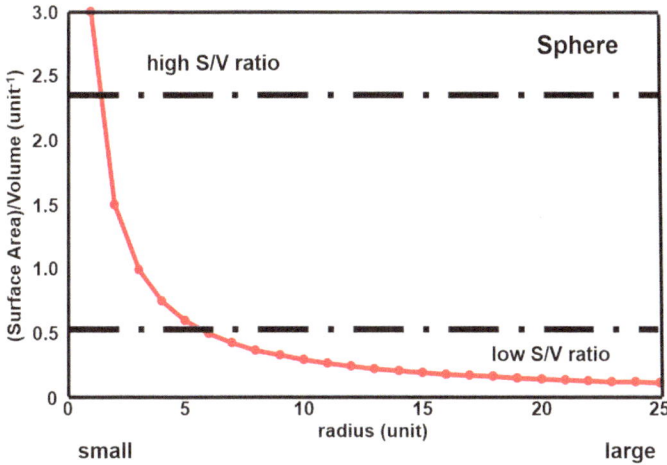

Fig. 2.7 Rapid decrease of (surface area)/volume ratio with increasing radius for a sphere

minced garlic would be 91 cm^2; because the total volume of the garlic is unchanged, the volume of minced garlic is still 2.3 cm^3, resulting in a total surface/volume ratio of ~ 40 cm^{-1} (identical to that of the individual minced garlic spherule). This is 400 times greater than the surface/volume ratio of the original clove. This enormous increase in the surface/volume ratio of the garlic results in a dramatic increase in the density of odor-producing molecules (of the volatile organic sulfur compound allicin) in the air. This increase produces garlic's characteristic pungent aroma.

Because Surface Area (SA) increases as the square of length while Volume (V) increases as the cube of length, we can write:

$$SA \propto V^{2/3}$$

Fig. 2.8 White garlic cloves, ready for cooking. *Image* by Jon Sullivan

where ∝ stands for "is proportional to". We can turn the proportion into an equation by inserting a constant, k:

$$SA = k \times V^{2/3}$$

The fraction 2/3 is approximately equivalent to the decimal 0.67; the equation can be rewritten as:

$$SA = k \times V^{0.67}$$

If we plot SA against V on a log–log diagram (e.g., Fig. 2.9), we get a regression line with a positive slope of 0.67 (where slope is defined as the change in the value of y divided by the change in the value of x; i.e., $\Delta y / \Delta x$).

If we are interested in the Surface Area per unit Volume, we can divide both sides of the equation by V:

$$SA/V = k \times V^{0.67}/V = k \times V^{0.67-1} = k \times V^{-0.33}$$

Plotting surface area per unit volume on a log–log diagram against volume yields a regression line with a negative slope of − 0.33 (e.g., Fig. 2.10).

If a collection of individual bodies (be they animals or asteroids) behaves in such a way that a log–log plot has a regression line of slope 0.67 or slope − 0.33, this would be consistent with these bodies having a geometric adherence to body surface area.

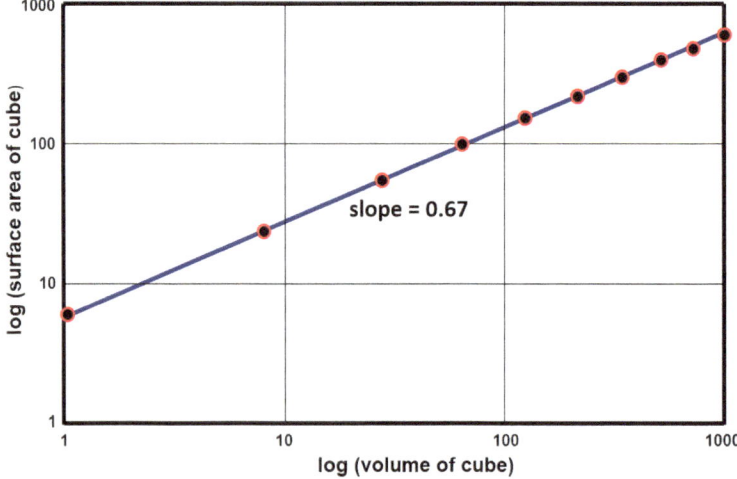

Fig. 2.9 Plot of surface area against volume for a cube on a log–log diagram, yielding a regression line with a slope of 2/3, equivalent to 0.67

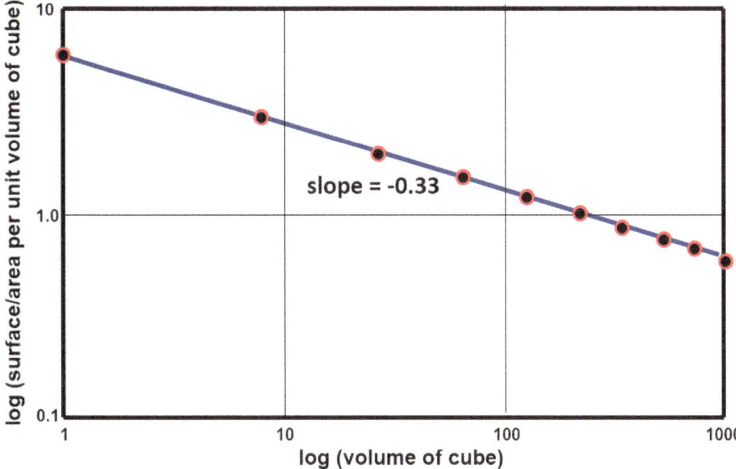

Fig. 2.10 Plot of surface area per unit volume versus volume for a cube on a log–log diagram. This yields a regression line with a negative slope: − 1/3 or − 0.33

That's it for the math. We can now explore how the surface/volume ratio affects the properties of a myriad of physical objects. We will understand why there are active volcanoes on Earth today, but not on the Moon. We'll find out why mice eat 15–20 times a day and why deciduous trees shed their leaves every Fall. We'll also find out why there is an average of more than 10 fatal agricultural dust explosions in the United States each year. These explorations are all based on simple geometric principles. As summarized by physicist and author Leonard Mlodinow: "Ever since the Greeks, mathematics has been at the heart of science, and geometry at the heart of mathematics."

Reference

El-Gayar, S. M., & Bahnas, O. T. (2005). Determination of physical and mechanical properties of garlic cloves. *MISR Journal of Agriculture Engineering, 22,* 427–439.

Chapter 3
Asteroids, Moons, Planets, and Meteorites

3.1 Inventory of the Solar System

Distances in the Solar System are often measured in astronomical units (abbreviated AU) where 1 AU is the mean distance between the Earth and the Sun. This is about 150 million kilometers, defined in 2012 as exactly 149,597,870.700 km. It is approximately 93 million miles or about eight light minutes. When we look at the setting Sun at dusk, we're seeing it as it was about eight minutes beforehand.

A cosmic accountant, eager to make an inventory of the Solar System, might create a ledger listing one Sun, eight major planets, about two hundred moons, a few dozen dwarf planets, more than a hundred-thousand asteroids larger than a kilometer, about ten million icy bodies in the Kuiper Belt (a torus extending from the orbit of Neptune at ~ 30 AU out to ~ 50 AU), and one or two trillion comets residing in the Oort Cloud (which extends an appreciable distance toward the nearest stars, up to ~ 200,000 AU).

The largest body in the Solar System is the Sun, a huge ball of gas 333,000 times more massive than the Earth. With a surface temperature of about 5500 °C, it floods the inner Solar System with heat and light. Fully 99.87% of the mass of the entire Solar System resides within the Sun; most of the rest is in Jupiter. The Sun consists of hydrogen (71.0% by mass) and helium (27.1%), with 0.97% oxygen, 0.40% carbon, and less than 0.1% of every other naturally occurring element. The size of the Sun is controlled by hydrostatic equilibrium – this is the balance between (1) an inward, compressive, pull of gravity (crushing everything inwards) and (2) an outward push arising from high internal gas pressure caused by the conversion of 600 billion kg of hydrogen into helium every second within the Sun's core.

The Sun is the center of the Solar System, the body around which all the planets revolve (Fig. 3.1). There are two basic types of planets:

The inner Solar System hosts the so-called "terrestrial planets", in reference to Earth, the largest of these bodies. These planets are composed mainly of metallic iron-nickel and silicate rock. Group members include Mercury (d = 4880 km), Venus (d = 12,104 km), Earth (d = 12,756 km), and Mars (d = 6794 km). Mercury is the

© The Author(s), under exclusive license to Springer Nature Switzerland AG 2023
A. E. Rubin, *Surface/Volume*, https://doi.org/10.1007/978-3-031-23749-2_3

Fig. 3.1 The Sun and its retinue of planets. The relative sizes of the bodies are accurate, but the distances among the bodies are not to scale

planet closest to the Sun with a mean distance of 0.387 AU. Moving outward, we encounter Venus (0.723 AU), Earth (1.000 AU, by definition), and Mars (1.524 AU). A martian year (686.98 days) is 1.88 times longer than an Earth year (365.256 days).

The outer Solar System is home to the "jovian planets", gas giants, named after the largest planet of all—Jupiter. These bodies include Jupiter (d = 142,984 km), Saturn (d = 120,536 km), Uranus (d = 51,118 km), and Neptune (d = 49,528 km). Jupiter and Saturn have dense atmospheres made mostly (> 99%) of hydrogen and helium. The interiors of these two planets consist of a large rocky core surrounded by a thick mantle of helium and liquid metallic hydrogen overlain by a shell of ordinary hydrogen and helium. Uranus and Neptune have atmospheres consisting mainly of hydrogen and helium along with some methane (CH_4). Their interiors consist of a large rocky core overlain by a mantle of highly compressed liquid water (with some dissolved ammonia, NH_3) surrounded by a spherical shell of liquid molecular hydrogen (H_2) and liquid helium. These four planets lie quite far from the Sun, underscoring the vastness of the Solar System: Jupiter (5.203 AU), Saturn (9.554 AU), Uranus (19.194 AU), Neptune (30.066 AU). While Jupiter orbits the Sun in just less than 12 years, it takes Neptune nearly 165 years.

Located between the orbits of Mars and Jupiter is the main asteroid belt, a zone about 1.21 AU wide, where upwards of 1 million asteroids reside (more than half of which have well-determined orbits and have been assigned official numbers). There are more than 100,000 bodies with diameters exceeding 2 km. They range in size from Ceres (d = 945 km), the largest asteroid (also considered a dwarf planet), down to bodies of meter and sub-meter size. Despite the large number of asteroids, their total mass is quite small, only about 3% the mass of the Moon.

We have samples of asteroids here on Earth. These are the meteorites, more than 99% of which were blasted off asteroids by high-velocity collisions of projectiles that were themselves derived mainly from fragmented asteroids. (The remaining meteorites in our collections hail from the Moon and Mars.)

Most asteroids were never appreciably heated; they retain basically the same minerals they had when they formed at the beginning of Solar-System history— around 4.567 billion years ago. Many of these asteroids are fairly homogeneous mixtures of abundant silicate minerals along with less-abundant grains of metallic iron-nickel and sulfide. These asteroids have about the same interelement proportions of non-volatile elements as the Sun's atmosphere.[1]

For example, the ratio of the bulk iron concentration to the bulk nickel concentration (i.e., Fe/Ni) throughout the interiors of unmelted asteroids is close to that in the Sun. The same is true for most other elements: e.g., Mg/Si, Ga/Ge, Cu/Al, Eu/Sm, Mo/Pb, Ir/Pd, V/Zn, etc..... Such asteroids are said to have "primitive" compositions because these bodies contain predominately unaltered materials formed from the first solids in the Solar System that agglomerated into rocks.

Before we discuss the other kind of asteroids—those that experienced wide-spread melting—we must stop at the kitchen counter. We are all familiar with the fact that water and oil don't mix; they are immiscible liquids, unable to dissolve in one another. If you take a glass of water, add some gold-colored vegetable oil, and shake it up, you'll see two separate liquids. The golden liquid with the lower density is the oil; it will accumulate at the top of the glass, floating on the water.

Just as oil and water mixtures are immiscible, so are mixtures of silicate liquids and metallic-iron-rich liquids. When a primitive asteroid (with homogeneously distributed silicates, metal grains, and sulfide grains) undergoes extensive melting, two immiscible liquids form: a low-density silicate liquid that floats to the top of the melted asteroid (to form the mantle and crust upon cooling), and a high-density metal-sulfide melt that sinks to the gravitational center of the body to form a metallic core.

Asteroidal or planetary bodies that experienced global melting are said to be "differentiated" because they develop a set of *different*, separated layers. From the center outwards, Earth is composed of an inner core of solid metallic iron-nickel, a surrounding spherical shell of liquid metallic iron-nickel, a thick solid silicate mantle, and a relatively thin silicate crust. [The core also contains a few percent lighter elements including S, O, Si, C, and H.] There are two kinds of crusts covering the

---

[1] Volatile elements are those that melt and vaporize at low temperatures. They include gases such as hydrogen and helium, as well as metals such as mercury and lead. Refractory elements melt and vaporize only at high temperatures. This group includes such metals as platinum and iridium, and elements such as aluminum and calcium that tend to enter silicate minerals and oxides.

Earth: an old 30–50-km-thick, lower-density granitic crust supporting the continents and a younger 5–10-km-thick, higher-density basaltic crust beneath the oceans.

Vesta, the second largest asteroid (d = 525 km) is also a differentiated body. It has a metallic iron-nickel core, a thick, silicate-rich mantle, and a basaltic crust. Computer simulations of impact cratering on Vesta suggest that the basaltic crust may be as thick as 80 km.

Beyond the eight major planets lies the Kuiper Belt. It is home to small bodies (Kuiper Belt Objects or KBOs) made mostly of frozen volatile compounds—ices of methane, ammonia, and water—and minor amounts of silicate rock. There are more than 100,000 bodies exceeding 100 km in diameter residing in the belt. The largest of these KBOs (as far as we know) is Pluto (d = 2377 km), ranked as a dwarf planet, along with Haumea (mean d = ~ 1650 km) and Makemake (mean d = ~ 1450 km). Neptune's large moon, Triton (d = 2707 km; ~ 14% larger than Pluto), may be a captured KBO.

The most distant part of the Solar System is the Oort Cloud, the theorized residence of a trillion or so icy planetesimals. The Oort Cloud is thought to consist of two zones—a disk-shaped inner cloud and a spherical outer cloud. Objects in the outer cloud have only loose gravitational ties to the Sun and can be easily unmoored by interactions with passing stars. Some of these small icy bodies could get diverted into new orbits that take them closer to the Sun. If they come close enough and begin to devolatilize, we see them as comets. All comets with orbital periods of 200 years or longer are thought to hail from the Oort Cloud.

3.2 Surface/Volume Effects in the Inner Solar System

There are six differentiated bodies of substantial size (hundreds to thousands of kilometers in diameter) in the inner Solar System that have experienced silicate volcanism: Mercury, Venus, Earth, Moon, Mars, and Vesta. A list of their geometric properties (Table 3.1) shows that the numerical values of their surface/volume ratios (km^{-1}) increase from Earth (4.70×10^{-4}) to Venus (4.96×10^{-4}) to Mars (8.83×10^{-4}) to Mercury (1.23×10^{-3}) to Moon (1.73×10^{-3}) to Vesta (1.14×10^{-2}).

All six bodies have volcanic rocks at their surfaces, indicating their mantles were heated sufficiently for such magmas to form. The most common volcanic rocks on Venus, Earth, Moon, Mars, and Vesta are basalts—fine-grained, dark-colored, low-viscosity silicate rocks dominated by two minerals: calcium-rich plagioclase (with a chemical formula in the range of $NaAlSi_3O_8$ to $CaAl_2Si_2O_8$, commonly also containing some potassium) and calcium-rich clinopyroxene ($Ca(Mg,Fe)Si_2O_6$). The volcanic rocks on Mercury are more magnesian than typical basalts and are considered intermediate in bulk chemical composition between basalt and komatiite (a variety of Mg-rich volcanic rock).

All bodies in the Solar System experienced at least some accretional heating. This was caused by the transformation into heat of the kinetic energy of projectiles striking the surfaces of these bodies during the early period in which these bodies agglomerated from smaller objects. After accretion, the planets were heated mainly

Table 3.1 Geometric properties of differentiated bodies in the inner Solar System

	Mercury	Venus	Earth	Moon	Mars	Vesta
Mean radius (km)	2439.7	6051.8	6371.0	1737.4	3389.5	262.7
Surface area (km^2)	7.48×10^7	4.60×10^8	5.10×10^8	3.79×10^7	1.44×10^8	8.66×10^5
Volume (km^3)	6.08×10^{10}	9.28×10^{11}	1.08×10^{12}	2.20×10^{10}	1.63×10^{11}	7.46×10^7
(Surface area)/volume (km^{-1})	1.23×10^{-3}	4.96×10^{-4}	4.70×10^{-4}	1.73×10^{-3}	8.83×10^{-4}	1.14×10^{-2}

by the decay of long-lived radioactive elements – isotopes of potassium (^{40}K) with a half-life ($t_{1/2}$) of 1.251 billion years, uranium (^{238}U; $t_{1/2} = 4.468$ billion years), and thorium (^{232}Th; $t_{1/2} = 14.05$ billion years). [With its shorter half-life ($t_{1/2} = 0.7038$ billion years) and low relative abundance (presently only 0.711% of natural uranium), ^{235}U was not a significant heat source early in Solar-System history.] Heat generated by the decay of long-lived radioactive isotopes accumulates slowly, so it is necessary for a body to be large (i.e., Moon- or planet-size) to retain the heat. Small bodies (with their high surface/volume ratios) lose heat too quickly to be heated in this way.

In contrast, the asteroid Vesta (which is far smaller than the Moon or any planet; Fig. 3.2) was likely heated predominantly by the decay of a radioactive isotope of aluminum (^{26}Al, with 13 protons and 13 neutrons) that has a much shorter half-life ($t_{1/2} = 717,000$ years). The decay product of ^{26}Al decay is ^{26}Mg (a stable isotope of magnesium with 12 protons and 14 neutrons). Excess amounts of ^{26}Mg have been found in basaltic meteorites called *eucrites* that likely originated on Vesta.

Paul Byrne (2020) used global maps of volcanic features of the Moon and inner planets to determine the approximate time that major volcanism ceased on these bodies:

Mercury—wide-spread extrusive volcanism lasted until about 3.5 billion years ago.

Venus—Although lava may have flowed as recently as 100,000 years ago (and might still be flowing), impact-crater statistics indicate that major global volcanism ceased about 750 million years ago.

Earth—Major volcanism continues today at divergent plate boundaries within the global mid-ocean ridge system. Volcanic eruptions occur daily; reports of current volcanic activity can be found at: https://www.volcanodiscovery.com/worldwide-volcano-activity/daily-reports.html.

Fig. 3.2 Relative sizes of the Earth (the largest differentiated body in the inner Solar System), the Moon, and the asteroid 4 Vesta (the smallest intact differentiated body in the inner Solar System)

Moon—More than 80% of the basalts on the dark, volcanic lunar plains (called *maria*, after the Latin word for seas) were emplaced by 2.5 billion years ago.

Mars—Major volcanism persisted in localized volcanic provinces until about 1.6 billion years ago.

We can also look at the asteroid Vesta, the smallest of these differentiated bodies. Radioactive dating shows that eucrites (basalts, most of which were probably derived from near-surface regions of Vesta) crystallized near the beginning of Solar-System history, about 4.55 billion years ago. The present average surface temperature at Vesta's equator is -123°C; lava no longer flows over the cold basalt.

Because volcanism persists today on Earth (the largest of these bodies) and petered out more than 4½ billion years ago on Vesta (the smallest of these bodies), it seems likely that the rate of global cooling of differentiated bodies is a function mainly of body size. There should be a rough correlation between the cessation of major volcanic activity and the surface/volume ratio of the differentiated bodies in the inner Solar System. This expectation is borne out (Fig. 3.3). All bodies except Mercury fall along or close to the curve. It appears that, *to a first approximation*, a differentiated parent body cools by radiation into space mainly as a function of its surface/volume ratio.

However, the real world is messy; there are often complexities and idiosyncrasies. The duration of volcanic activity is governed by several processes in addition to heat loss into interplanetary space; these include mantle bulk chemical composition, internal temperature, starting conditions, melt extraction, mantle convection, as well as the presence, absence, type, and vigor of plate tectonics. From Fig. 3.3, it appears that major volcanic activity on Mercury halted 1.5–2.0 Ga earlier than expected for a

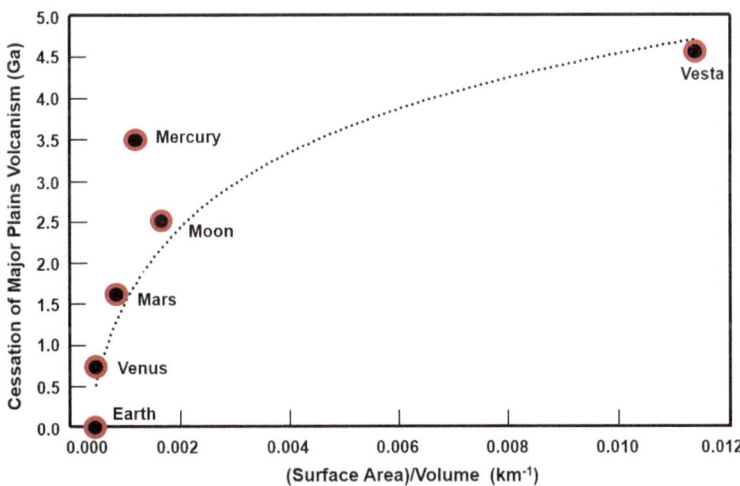

Fig. 3.3 The timing of cessation of major plains volcanism on differentiated bodies in the inner Solar System versus the (surface area)/volume ratio of those bodies. Mercury falls off the curve, indicating that a different property than the surface/volume ratio governs the cessation

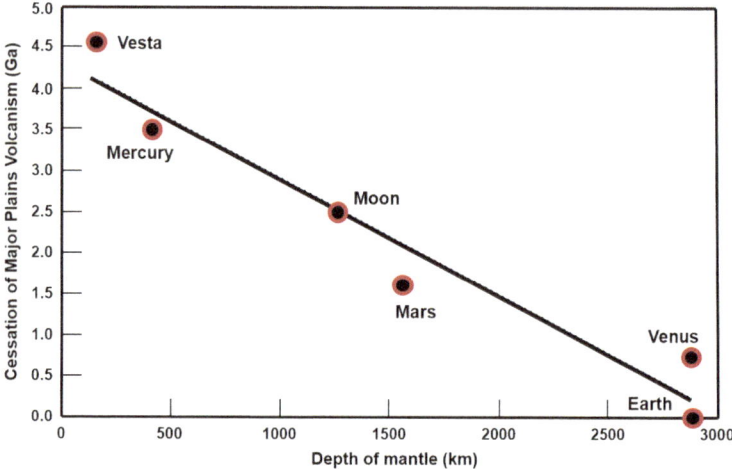

Fig. 3.4 Strong negative correlation between the timing of cessation of major plains volcanism on differentiated bodies in the inner Solar System and the total mantle depth (i.e., distance from the surface to the core-mantle boundary). Large bodies maintained volcanic activity for much longer periods

differentiated body with its surface/volume ratio. A different planetary characteristic than surface/volume must also significantly influence its rate of global cooling. This feature is total mantle depth (i.e., the total thickness of the silicate crust plus mantle); it is equivalent to the distance from the surface to the core-mantle boundary. Mercury has a large metallic iron-nickel core with a radius of 2024 km, equivalent to 83% of the planet's mean radius (2439.36 km). This indicates the volume of the core is 57% of the planet's volume. It also shows that the total depth of the mantle on Mercury (415 km) is much less than that of any other terrestrial planet.

If we plot the timing of the cessation of major volcanism on these bodies against the total mantle depth (Fig. 3.4), we get a very strong negative correlation, significant at the 99.86% confidence level.

Thus, the surface/volume ratio, although important, is not the sole determining factor in the cooling of large, differentiated bodies. Total mantle depth is also significant. Nevertheless, the smallest bodies—those with the largest surface/volume ratios—have the thinnest mantles. And that is why differentiated asteroids are presently colder than ice. They were heated by collisions and by the decay of ^{26}Al early in Solar-System history, but they radiated away their heat very quickly. The undifferentiated asteroids (those with primitive compositions that were never appreciably heated) also lost whatever heat they had very early in Solar-System history. Except for being heated by occasional collisions, all asteroids have been cold for more than 4½ billion years.

And the Earth—the largest planet in the inner Solar System (and the body with the lowest surface/volume ratio among terrestrial planets)—is still hot. Volcanoes erupt every day. There is a super volcano directly beneath Yellowstone National Park in the Western United States. It was responsible for super eruptions 2.1 million years ago, 1.3 million years ago, and 640,000 years ago. The next super eruption is likely to occur within a few hundred thousand years—the results would be catastrophic. As historian Will Durant noted in 1946: "Civilization exists by geological consent, subject to change without notice."

3.3 Meteorite Strewn Fields

And the second angel sounded, and as it were a great mountain burning with fire was cast into the sea… (Revelation 8:8; KJV)

Meteoroids enter the atmosphere from interplanetary space at high velocities, averaging about 18 km s^{-1}. They become luminous at altitudes of 80–90 km where the atmosphere is sufficiently dense for friction between the meteoroids and the molecules of nitrogen and oxygen to heat the surrounding air to incandescence. Depending on their friability and entry angle, large incoming meteoroids can begin to fragment at heights of ~ 50–60 km. For objects moving at 20 km s^{-1}, compressive stresses increase from 10 MPa (equivalent to the nozzle pressure of a high-pressure power washer) at an altitude of 30 km to 100 MPa at 15 km (Melosh, 1989). [The latter pressure is nearly 1000 times greater than atmospheric pressure at sea level.] This pressure exceeds the crushing strength of many stony meteorites. These objects may undergo a series of explosions, producing a hierarchy of fragmentation events. More than two-thirds of meteoroids fall into the ocean.

Individual fragments that survive and reach land produce an elliptical pattern known as a strewn field (e.g., Fig. 3.5). Strewn fields typically range from a few kilometers to tens of kilometers in length; most stony meteorite strewn fields are 2–4 km wide. The largest documented strewn field is that of the Gibeon IVA iron meteorite that fell in Namibia, probably in prehistoric times; iron fragments formed an ellipse ~ 100 km wide and 275 km long.

The most momentous recent fall, as of this writing, was that of the Chelyabinsk LL5 ordinary chondrite in Siberia. It was witnessed by thousands of people (and recorded on numerous dash cams) on 15 February 2015 at 9:22 AM local time. The 12,000–13,000 metric-ton meteoroid was ~ 20 m across; it entered the atmosphere over the southern Urals (Fig. 3.6), traveling roughly northwest at ~ 19 km s^{-1} at a shallow angle of ~ 18°. For a brief period, the meteoroid was brighter than the Sun.

The meteoroid exploded at an altitude of 29.7 km. It produced a shock wave with a total kinetic energy equivalent to about 25–30 Hiroshima atomic bombs. More than 7000 buildings were damaged from the shock wave. At least 1100 people were injured (mainly from shattered window glass) and sought medical attention.

On the order of 500,000 individual meteorites pelted the ground. Each sample is covered with fusion crust, a melt layer developed by frictional melting during atmospheric descent. The meteorites contain numerous shock melt veins (Fig. 3.7); these formed by small-scale impact events on the meteorite's parent asteroid long before Chelyabinsk's arrival on Earth.

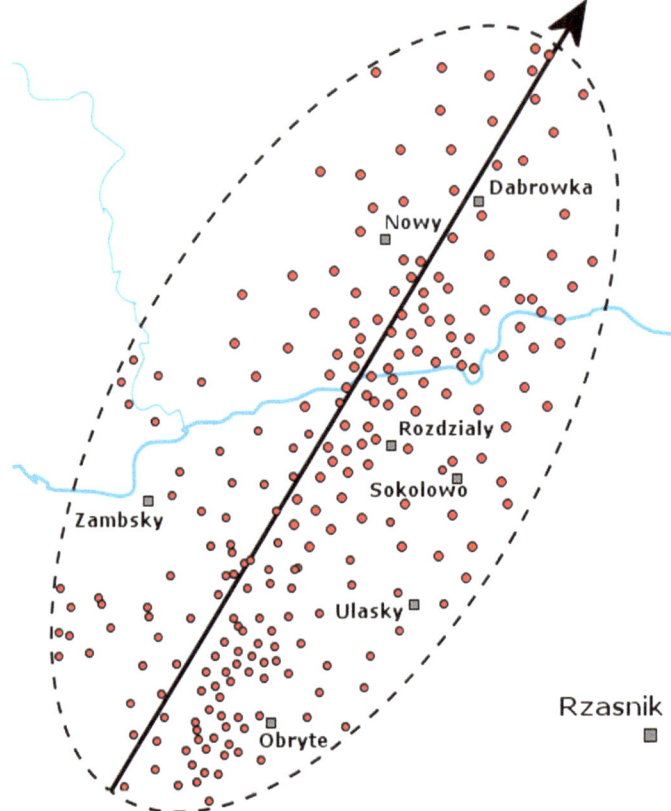

Fig. 3.5 Strewn field of the Pultusk H5 ordinary chondrite that fell in Poland in 1868. More than 100,000 pieces reached the ground. Each dot represents a recovered specimen. The arrow indicates the trajectory of the meteoroid. *Diagram* from Basilicofresco

The Chelyabinsk strewn field was about 1 km wide and 60 km long. The largest remaining fragment (the main mass) plunged into Lake Chebarkul, punching a 6-m-wide hole in the ice. When recovered, the mass weighed 654 kg. A plot of the distribution of the recovered masses (Fig. 3.8) shows a strong tendency for larger masses to occur farther downrange than smaller masses along the meteoroid's trajectory. The main mass sits at the far end of the strewn field; a dense concentration of small samples (<10 g each) occurs several kilometers to the southeast (in the direction from which the meteoroid came). Tens of thousands of sparsely scattered small samples (many less than 1 g) occur farther southeast along the trajectory.

The correlation between meteorite mass and location in strewn fields is a common one. The largest masses tend to occur near the terminus of the field (the direction in which the meteoroid was traveling); although there is appreciable overlap, small individuals tend to be recovered at the opposite end of the strewn field.

Why should this be so?

Fig. 3.6 Dust trail left by the Chelyabinsk fireball that exploded above the city of Chelyabinsk on 15 February 2013. This *photo*, taken by Alex Alishevskikh, was from a distance of ~ 200 km from the sub-meteoroid point

Fig. 3.7 Two specimens of the Chelyabinsk LL5 ordinary chondrite from the Muséum de Toulouse. *Photo* by Didier Descouens

Sample mass correlates with sample volume: for ideal, regularly shaped objects, mass and volume both increase as the cube of length. Thus, a low-mass sample also has a small volume.

Small samples (with low mass) have high surface/volume ratios. A spheroidal 1-g Chelyabinsk individual has a surface/volume ratio about 85 times greater than that of the main 654-kg mass. It is wind resistance (a.k.a. drag) that affects the sample distribution of incoming meteoroids. Small samples, with their high surface/volume ratios, are preferentially slowed by the wind. Larger samples retain more of their pre-atmosphere momentum (the product of mass and velocity) and travel the farthest; their small surface/volume ratios leave them less affected by drag. Nevertheless, the specific breakup dynamics of tumbling meteoroids during atmospheric passage complicate the distribution of individual fragments on the ground.

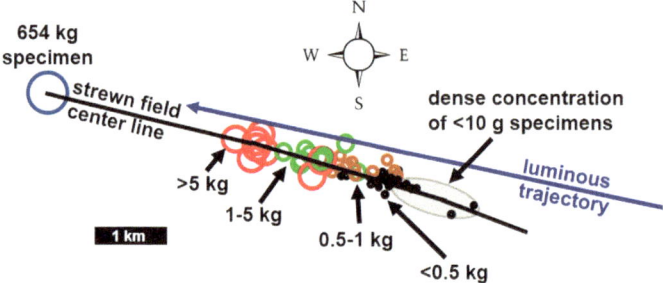

Fig. 3.8 Highly simplified map of the strewn field of the Chelyabinsk LL5 ordinary chondrite that exploded over Siberia on 15 February 2013. This map is based on a trajectory projection map by Svend Buhl (Buhl & Wimmer, 2013). Although there is significant overlap, individual meteorite specimens tend to be distributed according to their mass. The main mass (654 kg) is denoted by a blue circle; specimens > 5 kg, red; specimens 1–5 kg, green; specimens 0.5–1 kg, brown; specimens < 0.5 kg, black. The gray ellipse outlines a dense concentration of small (<10 g) specimens. The final portion of the luminous trajectory is shown; beyond the blue arrow, the meteoroid proceeded with dark flight until the main mass punched through the ice on Lake Chebarkul

Meteorite hunters plot the location of sample masses on terrain maps to identify the strewn field and aid their quest for large meteorites. Individual Chelyabinsk samples currently sell for $10–$20 per gram.

Air resistance (drag) can affect any object, not just an incoming meteoroid. As pointed out by the eminent Scottish geneticist, J. B. S. Haldane, in his 1926 essay, "On Being the Right Size": "You can drop a mouse down a thousand-yard [914-m] mine shaft; and, on arriving at the bottom it gets a slight shock and walks away, provided that the ground is fairly soft. A rat is killed, a man is broken, a horse splashes. For the resistance presented to movement by the air is proportional to the surface of the moving object. Divide an animal's length, breadth, and height each by ten; its weight is reduced to a thousandth, but its surface only a hundredth. So the resistance to falling in the case of the small animal is relatively ten times greater than the driving force." The moral lesson is that, when you have to throw an animal down a mine shaft (or off a cliff), choose a small one.

3.4 Chondrules

Let's move down in scale from the very large – asteroids, moons, and planets—down past the sizes of individual meteorites, and on to the very small—individual sub-millimeter-size components of these meteorites.

The most abundant, readily observable, component in most primitive, unmelted meteorites is the set of objects known as **chondrules** (pronounced kŏn' dr o͞o lz), etymologically derived from the ancient Greek word for "grain". The primitive meteorites that contain chondrules are called *chondrites*. Chondrules constitute 65–75% by volume of ordinary chondrites (the most common variety of meteorite observed to fall). The chondrules are made mainly of the magnesium- and iron-bearing silicate minerals olivine [(Mg,Fe)$_2$SiO$_4$] and low-calcium pyroxene [(Mg,Fe)SiO$_3$], along with silicate glass surrounding the olivine and pyroxene crystals. Many chondrules also contain small blebs of metallic iron-nickel (Fe–Ni) and Fe-sulfide.

The smallest chondrules are less than 1 μm in diameter (objects less than 40 μm across are known as "microchondrules") [where 1 μm is a thousandth of a millimeter]; the largest chondrules range up to about five centimeters in diameter. Typical chondrules are a few hundred micrometers across.

Just like snowflakes, all chondrules look different (Fig. 3.9). There are five common textural types: *porphyritic chondrules* have relatively large grains of olivine (PO) and/or pyroxene (POP or PP) surrounded by silicate glass; *barred olivine* (BO) *chondrules* are characterized by subparallel plates or bars of olivine separated by patches of glass; many BO chondrules are surrounded by a thin olivine spherical shell; *radial pyroxene* (RP) *chondrules* consist of laths of low-Ca pyroxene arranged in fan-like arrays radiating from a point near the chondrule surface; *cryptocrystalline* (C) *chondrules* are comprised of numerous patchy domains containing tiny low-Ca pyroxene grains and associated glass; *granular olivine-pyroxene* (GOP) *chondrules* consist of densely packed, small olivine and/or low-Ca pyroxene grains and small patches of glass.

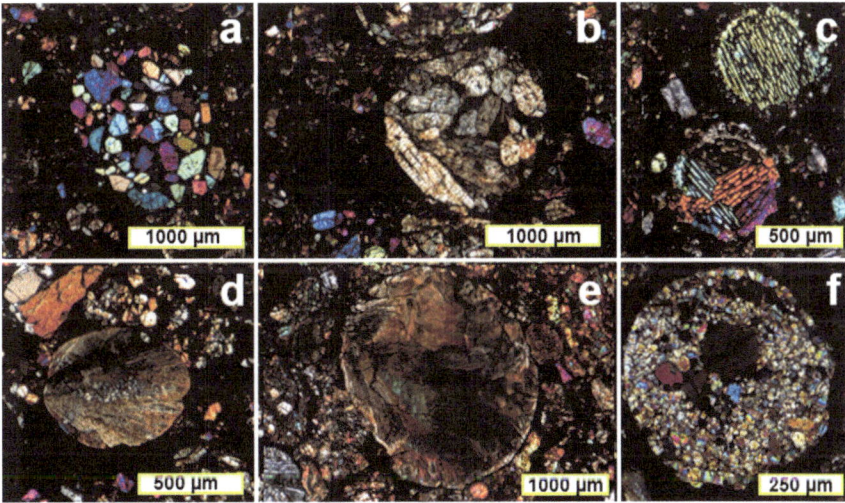

Fig. 3.9 Chondrules in ordinary chondrites. **a** Porphyritic olivine (PO) chondrule from an H3.8 chondrite from the Sahara (provisional designation Sahara 99228). **b** Porphyritic pyroxene (PP) chondrule from the LL3.1 chondrite NWA 2040. **c** Neighboring barred olivine (BO) chondrules (both with olivine shells) in the H4 chondrite NWA 4440. **d** Radial pyroxene (RP) chondrule from the H3.6 chondrite Clovis (no. 1). **e** Cryptocrystalline chondrule (exhibiting multiple domains) in the LL3.5 chondrite NWA 1096. **f** Granular olivine-pyroxene (GOP) chondrule from the H3.10 chondrite NWA 3358; the large grains near the center may be relict (incorporated from a preexisting chondrule). All images made using crossed polarizers. *Images courtesy* of John Kashuba

The round shapes of chondrules reflect their formation from isolated droplets of silicate melt. Due to the phenomenon of surface tension, liquid droplets have a strong tendency to shrink to the geometric form that has the minimum surface area for a given volume—that would be a sphere. Dewdrops, weld spatter, and tree resin are other examples of liquids that form spheroids because of surface tension.

Chondrules formed more than 4½ billion years ago from floating clumps of porous dust that orbited around the Sun before the era of planets. The clumps seem to have been heated and melted by short-lived bursts of energy. The source of this energy has not been identified, but one possibility is a large set of hypothetical lightning bolts generated in the solar nebula (cloud of gas and dust from which the Solar System sprang). Once the dust clumps melted, the proto-chondrule droplets cooled quickly to form the various textural types of chondrules we see today (Fig. 3.9).

Researchers have made synthetic chondrules in lab furnaces and inferred that real chondrules cooled over periods of seconds to tens of minutes. These workers found that porphyritic chondrules crystallize only from silicate liquids that contained small, unmelted, grain fragments (called nuclei) around which larger grains could nucleate and grow. In contrast, barred olivine (BO) and radial pyroxene (RP) chondrules formed typically from totally molten droplets free of tiny grain fragments.

Let's focus first on the BO chondrules. They range in size over three orders of magnitude. The small ones have much higher surface/volume ratios than the large ones and must have cooled much faster by radiating their heat away into space. Experiments show that the olivine bars in synthetic BO chondrules grow thicker at slower cooling rates. This suggests that small BO chondrules should have much thinner olivine bars than large BO chondrules. Is there evidence for this?

Rubin (1989) described a small rock fragment in the Krymka ordinary chondrite (a meteorite observed to fall in Ukraine in 1946) that contained olivine-rich microchondrules 3–31 μm in diameter. A few of the microchondrules have discernable BO textures and consist of subparallel 0.4- to 0.8-μm-wide olivine bars surrounded by glass. In contrast, typical BO chondrules in other meteorites are a few hundred micrometers across with olivine bars up to 100 μm wide. At the far end of the size spectrum are centimeter-size BO chondrules; the largest one reported is 1.3 cm in diameter. The outer portions of many normal- and large-size BO chondrules consist of spherical olivine shells (Fig. 3.9c); BO microchondrules lack such shells. Recent research shows that, within the set of BO chondrules in unequilibrated ordinary chondrites, there are significant positive correlations among chondrule diameter, bar thickness, and rim thickness. In the nebula, smaller BO precursor droplets cooled faster than larger droplets (due to their higher surface/volume ratios) and grew thinner bars and rims.

The surface/volume ratio of a 5-μm-wide BO microchondrule is 2600 times greater than that of a 1.3-cm-wide BO chondrule. This difference is responsible for the much faster cooling rates of the small BO chondrules and, consequently, the very thin olivine bars within them, and, probably also, the absence of spherical olivine shells.

Radial pyroxene (RP) chondrules also formed from totally molten droplets. Rubin et al. (1982) first identified RP microchondrules in the Piancaldoli ordinary chondrite (which fell in Italy in 1968). The smallest microchondrules were only 0.25 μm in diameter, observable only with a scanning electron microscope (SEM). The pyroxene laths in these RP microchondrules are very thin—some less than a few tenths of a micrometer wide.

The largest RP chondrules in meteorites are more than 4 mm in diameter, with pyroxene laths over 60 μm thick. The surface/volume ratio of a 4-mm-wide chondrule is 16,000 times smaller than that of a 0.25-μm-wide microchondrule. This geometric feature is responsible for large RP chondrules cooling much more slowly than small RP chondrules and, as a consequence, developing much thicker pyroxene laths during crystallization.

Rubin et al. (1982) measured the compositions of low-calcium pyroxene grains (a mineral of formula $(Mg,Fe)SiO_3$) in the centers of the RP microchondrules in the Piancaldoli meteorite. The compositions are expressed in mole% ferrosilite (abbreviated Fs), defined as the molar $(FeO)/(FeO + MgO)$ ratio of low-calcium pyroxene. The microchondrules, which contain little FeO and hence have low Fs values, are embedded in FeO-rich, fine-grained, silicate-rich, matrix material within the rim of a normal-size chondrule.

During the mild heating experienced by the Piancaldoli meteorite on its parent asteroid, diffusion caused iron atoms to move from the FeO-rich matrix into the microchondrules. [Diffusion is the process causing the iron atoms to move from a region of high iron concentration (the ferroan silicate matrix material) to a region of low iron concentration (the magnesian interiors of the microchondrules).] The smallest microchondrules have the highest surface/volume ratios; their centers are closer to the surface. One would expect the mobilized iron atoms to reach the centers of the smallest microchondrules more readily than the centers of larger microchondrules. That is exactly what was found (Fig. 3.10).

Fig. 3.10 Strong inverse correlation between the diameters of microchondrules and their central Fs contents in the Piancaldoli ordinary chondrite. *Diagram modified from Rubin et al. (1982)*

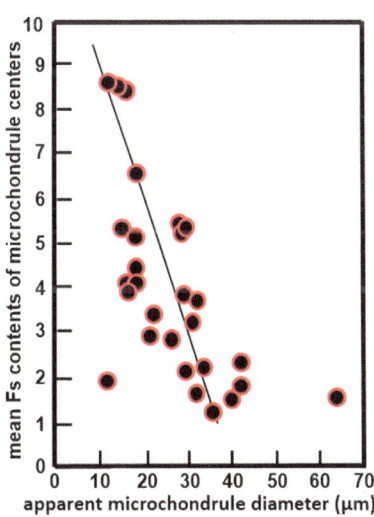

For the 28 analyzed microchondrules, the correlation coefficient $r = -0.60$, significant beyond the 99.9% confidence level. This means that the chance this is just a meaningless statistical fluke (instead of indicating that the correlation is real and robust) is about the same as someone flipping an honest coin ten times and getting all heads (1/1024):

$$\mathbf{H - H - H - H - H - H - H - H - H - H}$$

This example illustrates how the surface/volume ratio clearly governs the efficiency of diffusion in a set of similarly shaped objects.

3.5   Aqueous Alteration of Chondritic Meteorites

Phyllosilicates are water-bearing silicate minerals (e.g., clays) that have been identified in the spectra of undifferentiated asteroids. When these asteroids were heated early in Solar-System history (by the decay of the radioactive isotope ^{26}Al and/or by high-energy collisions), water within the asteroids was mobilized; it began reacting with anhydrous phases such as metallic Fe–Ni, Fe-sulfide, olivine [$(Mg,Fe)_2SiO_4$] and glass to produce new water-bearing minerals.

Chondritic meteorites were derived from undifferentiated asteroids, so it is not surprising that some ordinary and carbonaceous chondrites contain phyllosilicates as well as other minerals formed during parent-body aqueous alteration.

The principal physical components of chondrites are chondrules, calcium-aluminum-rich inclusions (CAIs) [submillimeter-to-centimeter-size objects composed of refractory minerals that formed at high temperatures, a.k.a. refractory inclusions], metallic Fe–Ni, Fe-sulfide, and fine-grained silicate-rich matrix material. When water was mobilized during heating, the fine-grained matrix material altered first. The small grains in the matrix (typically < 1–5 μm in size) have very high surface/volume ratios and are quite susceptible to alteration. Chemical reactions between water and these fine silicate grains readily affected the entire particles. The CI-group of carbonaceous chondrites is very fine-grained and extensively altered; water-bearing silicates constitute ~ 95 vol.% of these meteorites.

During aqueous alteration of some carbonaceous and ordinary chondrites, metallic Fe–Ni grains within them were oxidized and converted into magnetite (Fe_3O_4). Oxidation occurred at grain surfaces. Larger grains developed thin rinds; intermediate-size grains developed high magnetite/metal ratios; and smaller grains were completely oxidized (Fig. 3.11).

Another example of aqueous alteration in chondritic meteorites was described by Grossman et al. (2000). They examined "bleached" radial pyroxene (RP) and cryptocrystalline (C) chondrules in ordinary chondrites and found that aqueous fluids had dissolved interstitial glass in the outer zones of these chondrules. Chondrule glass was partially replaced by phyllosilicate, leaving the outer ~ 100 μm zones lighter in color ("bleached"). Because smaller chondrules have higher surface/volume ratios,

Fig. 3.11 Schematic diagram of metallic Fe–Ni grains of different sizes converted to magnetite during parent-body aqueous alteration

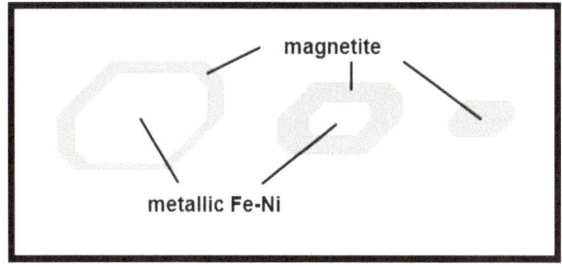

Fig. 3.12 Response of cryptocrystalline chondrules of different sizes to bleaching. Smaller chondrules have higher proportions of bleached zones; very small chondrules can be entirely replaced by alteration products

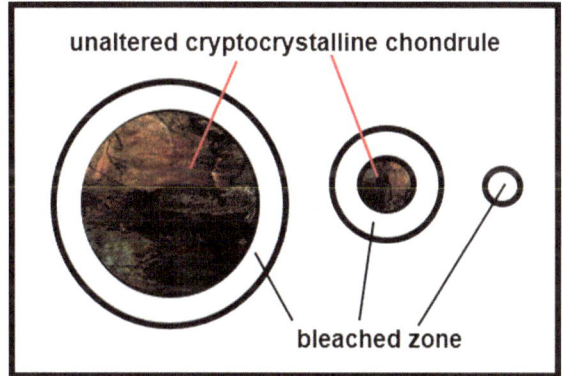

higher proportions of these chondrules became bleached during episodes of alteration (Fig. 3.12). Some very small cryptocrystalline chondrules were entirely replaced by phyllosilicates and other alteration phases.

Carbonaceous chondrites inherited components with different oxygen-isotopic compositions from the solar nebula. Different mineral grains in fine-grained CAIs in the more-altered members of the CV and CO groups of carbonaceous chondrites have heterogeneous oxygen-isotopic compositions. The grains are typically 10–20 μm in size. Oxygen has three stable isotopes with different numbers of neutrons. The most abundant isotope is ^{16}O with eight protons and eight neutrons. Some minerals in the fine-grained CAIs (e.g., anorthite—$CaAl_2Si_2O_8$, melilite—$(Ca,Na)_2(Al,Mg)(Si,Al)_2O_7$, grossite—$CaAl_4O_7$, perovskite—$CaTiO_3$) are depleted in ^{16}O relative to other minerals (e.g., corundum—Al_2O_3, hibonite—$CaAl_{12}O_{19}$, spinel—$MgAl_2O_4$, forsterite—Mg_2SiO_4) (Krot, 2019). The diffusion rate of oxygen is higher in the ^{16}O-depleted minerals; their present O-isotopic compositions were set by reaction with ^{16}O-poor aqueous fluids during parent-body alteration. Because the mineral crystals in these fine-grained CAIs were so small (and had very high surface/volume ratios) the grains altered rapidly after water in the asteroids was mobilized during heating.

3.6   Cosmic Spherules

Each year, roughly 40,000 metric tons of cosmic dust hit the top of the Earth's atmosphere; about 90% of this material is vaporized or disaggregated into submicrometer, smoke-size dust grains. Surviving materials that reach Earth's surface as 1-μm- to 2-mm-size particles are known as micrometeorites. About 71% of micrometeorites fall into the ocean. The precursors of the micrometeorites enter the Earth's atmosphere at cosmic velocity, typically ~ 18 km s^{-1}, where they undergo extreme heating due to friction with the surrounding air molecules. Although some particles traverse the atmosphere and remain largely unmelted (and retain their initial mineralogy), other particles are partly melted and contain a few unmelted grains. Many particles melt completely; these liquid droplets (which assume spheroidal shapes due to surface tension) are called cosmic spherules. Some spherules cool rapidly during their descent and transform into glass; others cool more slowly and develop igneous textures similar to those of many chondrules, e.g., porphyritic, barred olivine, and cryptocrystalline (Fig. 3.13).

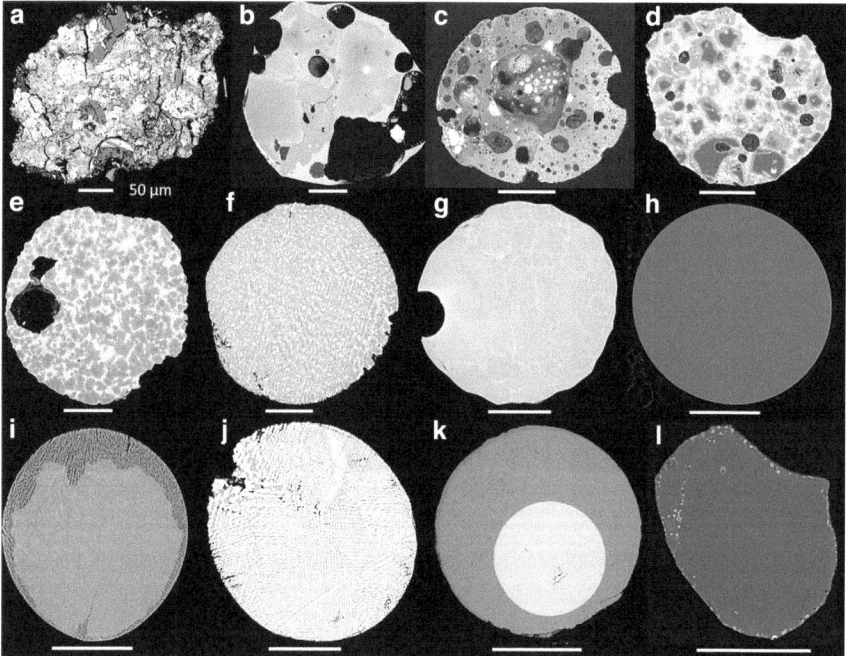

Fig. 3.13 Cross sections of different types of micrometeorites (a.k.a. cosmic spherules). **a** Fine-grained unmelted; **b** Coarse-grained Unmelted; **c** Scoriaceous; **d** Relict-grain bearing; e Porphyritic; **f** Barred olivine; **g** Cryptocrystalline; **h** Glass; **i** CAT; **j** G-type; **k** I-type; and **l** Single mineral. All varieties are silicate rich except for G- and I-types. All scale bars are 50 μm long. *Image courtesy* of Susan Taylor, obtained from Wikimedia Commons, authored by Shaw Street

Cosmic spherules were first collected from the ocean floor during the *Challenger* expedition (1872–1876) by the *HMS Challenger*, a small British warship converted into a scientific vessel by the Royal Navy. The ship covered about 70,000 nautical miles (130,000 km) in a rigorous program of exploration, dredging and surveying. The 1891 scientific report by Murray and Renard announced the discovery of two types of micrometeorites: black magnetic spherules (some with relict metallic Fe–Ni grains) and brown-colored silicate spherules with chondrule-like textures. A couple of years later, these researchers concluded the spherules were extraterrestrial because (a) they were recovered far from potential terrestrial sources, (b) they were most prevalent in slowly accumulated sediments, (c) their textures were unlike furnace products, (d) their metallic Fe–Ni grains differed from those in volcanic rocks, and (e) some of the larger spherules contained platinum-group elements.

Clayton et al. (1986) examined a suite of deep-sea cosmic spherules sorted into different size fractions. They separated metallic-iron-bearing spherules (Type I) from silicate spherules (Type S) and determined their oxygen-isotopic compositions. The Type I spherules contained some bulk Ni and consisted of magnetite (Fe_3O_4) and wüstite (FeO); some had relict cores of metallic Fe–Ni. Because these iron-oxide minerals formed from oxygen-free metallic Fe–Ni during atmospheric descent, all the oxygen within them initially came from the upper reaches of the Earth's atmosphere (at altitudes of 80–120 km).

These atmospheric regions appear to have high $^{17}O/^{16}O$ and $^{18}O/^{16}O$ ratios relative to the oxygen-isotopic compositions of materials on or near the Earth's surface. However, after residing in the ocean for long stretches of time, some of the atmospheric oxygen exchanged with oxygen atoms in sea water (which has a different oxygen-isotopic composition). It appears that the smallest Type I spherules (which have the highest surface/volume ratios) were altered by seawater to a greater extent than the largest spherules. Specifically, the small spherules (100–150 μm in diameter), with a mean surface/volume ratio of 0.048 μm^{-1}, contained ~ 15 wt.% more alteration products than the larger (400 μm) spherules, with a mean surface/volume ratio of 0.015 μm^{-1}. The seawater was able to exchange O isotopes more readily with the interiors of the smaller cosmic spherules because their interiors were closer to the spherule surface (i.e., the spherule/seawater boundary).

References

Buhl, S., & Wimmer, K. (2013). Trajectory projection of the Chelyabinsk superbolide and location of recorded meteorite finds. OpenStreetMap contributors (openstreetmap.org), www.meteorite-recon.com [map].

Byrne, P. K. (2020). A comparison of inner Solar System volcanism. *Nature Astronomy, 4*, 321–327.

Clayton, R. N., Mayeda, T. K., & Brownlee, D. E. (1986). Oxygen isotopes in deep-sea spherules. *Earth and Planetary Science Letters, 79*, 235–240.

Grossman, J. N., Alexander, C. M. O. 'D., & Brearley, A. J. (2000). Bleached chondrules: Evidence for widespread aqueous processes on the parent asteroids of ordinary chondrites. *Meteoritics & Planetary Science, 35*, 467–486

Krot, A. N. (2019). Refractory inclusions in carbonaceous chondrites: Records of early solar system processes. *Meteoritics & Planetary Science, 54*, 1647–1691.

Melosh, H. J. (1989). *Impact cratering: A geologic process*. Oxford University Press, 245 pp.

Rubin, A. E. (1989). An olivine-microchondrule-bearing clast in the Krymka meteorite. *Meteoritics, 24*, 191–192.

Rubin, A. E., Scott, E. R. D., & Keil, K. (1982). Microchondrule-bearing clast in the Piancaldoli LL3 meteorite: A new kind of type 3 chondrite and its relevance to the history of chondrules. *Geochimica et Cosmochimica Acta, 46*, 1763–1776.

Williams T. (1953). Camino Real. In M. Gussow & K. Holditch (Eds.), Williams, Tennessee. Plays 1937–1955. Library of America, 2000, New York.

Chapter 4
Geologic Processes

Truth shall spring out of the Earth... (Psalm 85:11; KJV)

4.1 Extrusive and Intrusive Igneous Rocks

The crust of the Earth is composed of three principal types of rocks:

Sedimentary rocks formed from sediments that were deposited grain-by-grain or rock-by-rock in a fluid medium (i.e., water or air) or precipitated from saturated watery solutions. Sediments that have been transformed into sedimentary rocks include sands deposited in shallow ocean water or blown into dunes by the wind. Deep burial of these sediments caused them to lithify into sandstone. Similarly, mud (rich in clay minerals) deposited in deep ocean water can be buried and transformed by mild heat and pressure (a process called diagenesis) into shale. Conglomerates are sedimentary rocks formed from lithified deposits of gravel, derived from riverbeds. Coal is a biologically derived sedimentary rock produced after decaying plant matter in fresh-water swamps formed peat; after burial, peat was transformed by heat and pressure into coal—a black combustible rock. Calcium carbonate ($CaCO_3$) can precipitate in warm shallow ocean water or in fresh-water lakes to form chemical limestones. More common than chemical limestones are biological limestones, many of which contain fossils; these limestones formed by the slow accumulation at the sea floor of coral debris and/or the calcium-carbonate shells of foraminifera (single-celled marine organisms).

Metamorphic rocks are made from preexisting sedimentary, metamorphic, or igneous rocks by some combination of significant heat, pressure, and hot mineral-rich fluids. Metamorphism could occur by burial deep within the Earth's crust or by the magmatic intrusion at shallower depths of a sill sandwiched between layers of sedimentary rock or of a dike pushing its way at a random angle though preexisting rock. Metamorphism causes significant chemical and/or physical changes in

A. E. Rubin, *Surface/Volume*, https://doi.org/10.1007/978-3-031-23749-2_4

Fig. 4.1 A chunk of
obsidian exhibiting
conchoidal fracture. Image
from Ji-Elle

the affected rocks. Common metamorphic rocks include quartzite (metamorphosed quartz-rich sandstone or chert), slate (metamorphosed shale), schist (metamorphosed slate), marble (metamorphosed limestone or dolomite) and gneiss (some samples of which are metamorphosed granites). Glacial ice, made of frozen H_2O, is commonly considered a metamorphic rock, formed after moderately high pressures compressed sedimentary layers of snow.

<u>Igneous rocks</u> formed from a cooling melt or a melt-crystal mush. Slow cooling of this liquid led to eventual crystallization of mineral grains; quenching (very rapid cooling) of the melt led to the formation of glass, e.g., tachylite (a black or dark-brown, low-silica, brittle, basaltic volcanic glass) or obsidian (a black, silica-rich volcanic glass; Fig. 4.1).

The most common volcanic rock is basalt (Fig. 4.2a); it constitutes more than 90% of all volcanic rocks on Earth. Basalt also occurs on the Moon, Mars, Venus, and the asteroid Vesta; the volcanic rocks on Mercury may be somewhat richer in MgO and poorer in K_2O than normal basalt.

Basalt is a fine-grained, dark-colored rock, extruded on Earth as low-viscosity lava from shield volcanoes in such places as Hawaii and Iceland; it also flows from fissures on the ocean floor. Andesite is a fine-grained volcanic rock with a bulk composition richer in silica than basalt. The most siliceous volcanic rock is called rhyolite (Fig. 4.3a); it is light-colored, generally pinkish, and typically fine grained. Because rhyolitic lavas are so viscous, pressure can build up in their magma chambers, resulting in explosive volcanic eruptions.

In contrast to fine-grained volcanic rocks formed at or near the Earth's surface, plutonic rocks are coarse-grained samples formed in vast magma chambers deeper within the Earth's crust. These rocks crystallized slowly, typically over tens of thousands of years. The most common plutonic rock is granite (Fig. 4.3b), a rock that has the same chemical composition as rhyolite. Batholiths are large masses of granitic rocks with areal extents of at least 100 km^2. Perhaps the most famous example is the Sierra Nevada Batholith, lying at the core of the Sierra Nevada Mountains in California.

Fig. 4.2 Rocks of basaltic bulk composition. **a** An unusual olivine-plagioclase phyric basalt from a drill core at Western Cape Cod, Massachusetts. The sample contains 60% plagioclase, 25% Ca-pyroxene, 8% chlorite and 7% oxide minerals. The original olivine crystals were replaced by chlorite during alteration after the basalt was erupted. The large, light-colored phases are also chlorite, formed by alteration of coarse olivine phenocrysts. Image courtesy of the U.S. Geological Survey. **b** Gabbro from Rock Creek Canyon, eastern Sierra Nevada, California. Image modified from that of Mark A. Wilson

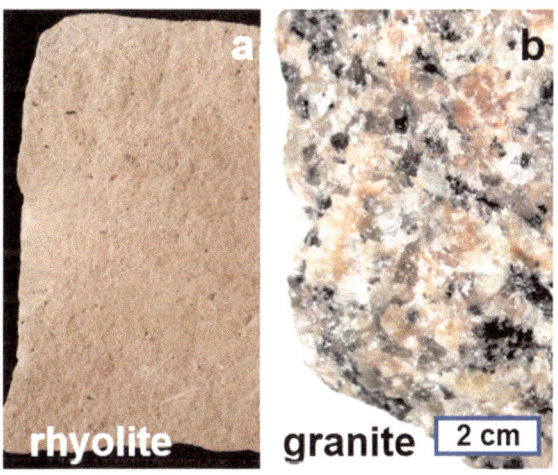

Fig. 4.3 Rocks of granitic bulk composition. **a** Rhyolite. Image from Michael C. Rygel via Wikimedia Commons. **b** Granite. Coarse-grained rock composed of major quartz (white), potassium feldpar (pink) and biotite (black)

Gabbros are also plutonic rocks (Fig. 4.2b); they are dark-colored, coarse-grained rocks with the same bulk chemical composition as basalt. They formed at depth in the crust from slowly cooling magma. A large fraction of the ocean crust is composed of gabbroic rocks.

Table 4.1 Similarities in mean bulk chemical composition between basalts and gabbros

	Tholeiitic basalt	Gabbro
No. of analyses	137	160
SiO_2	50.83	48.36
TiO_2	2.03	1.32
Al_2O_3	14.07	16.84
Fe_2O_3	2.88	2.55
FeO	9.00	7.92
MnO	0.18	0.18
MgO	6.34	8.06
CaO	10.42	11.07
Na_2O	2.23	2.26
K_2O	0.82	0.56
H_2O	0.91	0.64
P_2O_5	0.23	0.24
Total	99.94	100.00

Tholeiitic basalt is the most common type of basalt; it formed at mid-ocean ridges. Data from Table 4.1 of Mueller and Saxena (1977)

For illustrative purposes, we will focus on basalts and gabbros. These rocks have the same bulk chemical composition, typically containing 45–52 wt.% SiO_2, 2–5 wt.% alkalis (i.e., Na_2O + K_2O), 0.5–2.1 wt.% TiO_2, 5–14 wt.% FeO, and at least 14 wt.% Al_2O_3. The close compositional similarities between basalt and gabbro are shown in Table 4.1. There are always minor variations in bulk composition between different rocks of the same type; thus, it is to be expected that the average bulk compositions of basalts and gabbros would not be identical.

Gabbros are medium-to-coarse-grained igneous rocks, with grain sizes ranging from 1–2 mm to 2–3 cm, depending on the specimen. They are composed mainly of calcium pyroxene and calcium plagioclase (in subequal amounts) with minor-to-accessory amounts of hornblende, olivine, and other phases. Basalts are fine-grained rocks exhibiting a variety of textures and grain sizes: some are glassy; others are cryptocrystalline with average grain sizes less than a few hundred micrometers; many are porphyritic with millimeter-size crystals of Ca-plagioclase ± Ca-pyroxene. Some basalts contain a few coarse grains (xenocrysts) or rock fragments that crystallized before the basalt erupted and were carried to the surface by ascending basaltic magma.

If basalts and gabbros have essentially the same bulk chemistry and mineralogy, why do they differ so much in average grain size? This difference is due to the rate of cooling from high temperatures (at which point the material is completely molten) down to the temperatures at which mineral grains begin to nucleate and crystallize. Basalts cooled rapidly, roughly on the order of 2 °C/hr (Lofgren, 1979, 1983); gabbros cooled much more slowly, mainly in the range of 10^2 °C/Ma to 10^4 °C/Ma (i.e., about

Fig. 4.4 Extruded basalt flows. **a** Basaltic lava (red) visible through a collapsed roof in solidified pahoehoe (ropy) lava at the Kilauea volcano in Hawaii. Image courtesy of the U.S. Geological Survey. **b** Thin flow of basaltic lava exuded from the Kilauea volcano, burning asphalt as it moves across Chain of Craters Road in Hawaii. Photo taken on 13 February 2003 from Hawaiian Volcano Observatory

a million to a hundred million times slower than basalt) (Gee & Meurer, 2002; Usui, 2013).

Many basalt lava flows are 1–10 m thick (Fig. 4.4a, b). Ropy pahoehoe lava flows (Fig. 4.4a) are thinner—typically only a few tens of centimeters thick. The tholeiitic flood basalts of the Columbia River Basalt Group extend from northern Oregon to central and eastern Washington to western Idaho, covering 2×10^5 km^2 of the landscape (Waters, 1962). Individual lava flows in this group range from less than a few meters up to 45 m in thickness. Flow volumes are typically 10–20 km^3 (BVSP, 1981).

For purposes of calculation, we will assume a 5-m-thick flow with a total volume of 15 km^3 (the center of the range). A five-meter-tall slab in the form of a right rectangular prism with this same volume would have dimensions of 54.8 km $\times$ 54.8 km $\times$ 0.005 km. Its surface area would be ~ 6000 km^2 and its surface/volume ratio would be ~ 400 km^{-1}.

Gabbros crystallized within large magma chambers with radii on the order of 1–7 km (Blake, 1981). If we assume a spherical magma chamber with the same volume as the estimated average Columbia flood basalt flow (15 km^3), the radius of the chamber would be about 1.5 km (well within the observed range of such chambers). The surface area of this magma chamber would be about 28 km^2 and the surface/volume ratio would be ~ 1.9 km^{-1}.

This calculation shows that the surface/volume ratio of the extrusive basaltic slab is more than 200 times greater than that of the gabbroic magma chamber. That dramatic difference in surface/volume ratio accounts for much of the difference in cooling rate (and consequently in grain size) between the fine-grained basalts (rapid cooling rates; high surface/volume ratios) and the coarse-grained gabbros (slow cooling rates; low surface/volume ratios). An additional significant cooling effect for the basalts is due to its location at the Earth's surface—one side of the extrusive flow is cooling radiatively into the air. In contrast, all sides of the gabbroic magma chamber are insulated by massive silicate rock (which has a much lower thermal diffusivity than air).

An analogous relationship applies to rhyolites and granites. Extrusive flows of fine-grained, rapidly cooled rhyolites have much higher surface/volume ratios than the large magma chambers of slowly cooled, coarse-grained granites.

4.2 Permeability of Sandstones

Crude oil and natural gas occur commonly within permeable rock layers (e.g., sandstone) sandwiched between impermeable layers (e.g., shale). These fuel sources can be trapped in localized regions below ground by folds and faults (structural traps) or by discontinuous layers of sedimentary rocks (stratigraphic traps). Boreholes drilled at these sites are used to explore potential reserves; vertical pipes can be inserted into boreholes to examine the different subsurface rock layers when the pipes are extracted.

A few definitions:

Porosity, symbolized by the Greek letter *phi*, Φ, is the proportion of void space in a material, whether it is a rock, an asteroid, or a fabric. It is the ratio of the volume of void space to the total volume ($\Phi = V_{voids}/V_{total}$). Igneous rocks such as granite and basalt are made of interlocking mineral grains (some also contain glass); these rocks have very low porosities, typically $< 1\%$. The only pores present are likely to be fractures. Sandstones have much higher porosities, commonly 10–35%; the mineral grains are not fitted tightly together. Fractures may also be present.

Permeability, symbolized by k, is a fundamental property of porous substances. This parameter is defined as the capacity of a fluid to be transmitted through a material. Although permeability is a function of bulk porosity, it is affected by pore size, pore shape, pore interconnectedness, and fluid pressure. It also depends on the molecular size of the fluid (e.g., helium flows more easily than water) as well as the fluid's viscosity (and viscosity is, to a significant extent, a function of temperature). The common standard unit of permeability is the millidarcy (md), approximately equivalent to 10^{-9} m^2. Unconsolidated sediments (e.g., sand) have very high permeabilities, in many cases exceeding 5000 md. Lithified sedimentary rocks of relatively high permeability, such as many sandstones, permit significant fluid flow; sedimentary rocks of low permeability, such as shale, inhibit fluid flow. Rocks with permeabilities greater than 100 md allow petroleum extraction without hydraulic stimulation such as fracking. Rocks with lower permeabilities can form stratigraphic seals and, under favorable geologic conditions, serve as petroleum traps.

Tortuosity is an indicator of the convoluted (tortuous) pathway a fluid moves through a material. It is defined as the ratio of the length of the actual fluid pathway between two given points to that of a straight-line segment connecting those same points. Such a tortuosity factor can be symbolized by Φ^m (where Φ is the porosity and m is the conductivity exponent). ($\Phi^m = L_{actualpath}/L_{straightline}$).

For many sandstones, the surface/volume ratio is controlled by the clay particles within them. For example, a coarse sandstone may contain quartz grains averaging 1 mm (or 1000 μm) across, whereas clay particles in the same rock may average 1 μm. The surface/volume ratios of spherical particles with these dimensions differ by a factor of 1000: i.e., 0.006 (sand-size quartz grains) versus 6 (clay).

The permeability of the rock is also affected by clay particles because they can clog up pores.

Sen et al. (1990) examined 126 sandstone samples from around the globe, mainly from oil-producing sites. The samples ranged in permeability over more than five orders of magnitude (from 0.01 to 3553 millidarcys), reflecting differences in grain size, grain density, and abundance of clay minerals.

These researchers measured various properties of these sandstones including:

(1) The pore volume (V_p) per unit surface area (S), a ratio denoted as (V_p/S); this is the reciprocal of the surface/volume ratio for the pores.
(2) The permeability (k).

There is a strong positive correlation (Fig. 4.5) between the log of the permeability of the sandstones [i.e., log (k)] and the log of a parameter that combines the pore volume per unit surface area with the tortuosity factor [i.e., log ($\Phi^m V_p/S$)].

Fig. 4.5 Diagram showing the strong correlation between the logarithm of the permeability (k) versus the logarithm of the pore volume per unit surface area combined with the tortuosity factor ($\Phi^m V_p/S$) Modified from Sen et al. (1990)

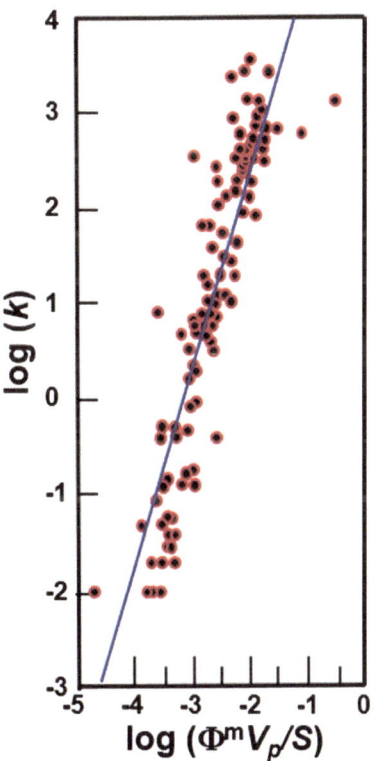

Fig. 4.6 Idealized cartoon
of a thin section of a poorly
sorted, feldspar-rich
sandstone with abundant
pore space (blue). Small clay
grains occur within the pores

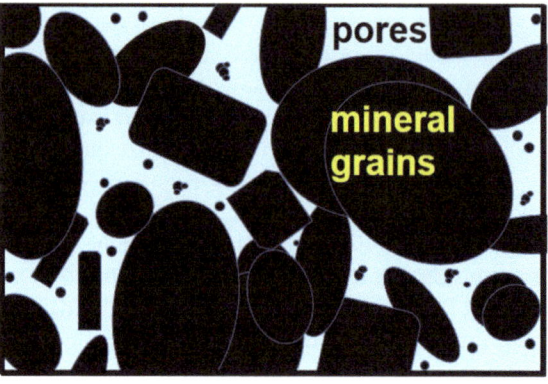

For illustrative purposes, we could omit the tortuosity factor on the horizontal axis
of this figure (i.e., the x-axis or abscissa) and use the reciprocal of the volume/surface
ratio to obtain $\log (S/V_p)$ as the unit for the x-axis. In that case, we would get a strong
negative correlation between permeability and the pore surface/volume ratio. In other
words, as the surface/volume ratio increases, the permeability decreases. Because
clay particles are tiny, each particle has a large surface/volume ratio. Quartz-rich
sandstones (which have high quartz/feldspar ratios) can contain up to 15 vol.% clay
minerals; wackes (sandstones with lower quartz/feldspar ratios) contain *more* than
15 vol.% clay. A sandstone containing abundant clay would have a microstructure
with a high surface/volume ratio (e.g., Fig. 4.6). Because clay particles clog pores,
they reduce the pore volume and decrease the permeability of the whole rock. Thus,
the surface/volume ratio of a sandstone microstructure directly affects the rock's
permeability: high S/V means low permeability.

4.3 Erosion and Weathering

Erosion is the set of processes involved in breaking down and lowering topographic
highs; soil and rock are loosened from their initial sites and transported downhill or
downwind. Steep slopes, especially those lacking vegetation, are more likely than
shallow slopes to experience landslides and mudslides during periods of heavy rain.
Erosion is nature's attempt to turn mountains into molehills.

Deposition is the set of processes responsible for filling in topographic lows, e.g.,
the infilling of basins with wind-blown silt or with sediments laid down in lakes or
ponds.

Taken together, the results of erosion and sedimentation tend to make the Earth
flatter. These forces are countered by: (1) plate tectonics causing (a) mountain
building at convergent boundaries, (b) volcano construction downrange of subducting
slabs at convergent boundaries, and (c) volcanic ridges at divergent-boundary

spreading centers, as well as (2) isostatic rebound (the gradual uplift of land masses previously depressed by the heavy load of kilometer-thick glaciers).

Weathering is a localized process, acting to break individual rocks apart. A rock cropping out at the Earth's surface is stuck between a rock and a hard place. It may be exposed to wind, acid rain, snow, and hail; it could endure (a) abrasion from sandblasting and water torrents, (b) pounding by waves, (c) abrasion and scouring by moving glaciers, or (d) crushing by falling rocks. Pebbles can be swallowed by animals that lack grinding teeth, serving as gastrointestinal aids in grinding food. [These animal ingurgitators include crocodiles, alligators, ostriches, sea lions, and seals; the swallowed stones are known as gastroliths.]

There are two principal modes of weathering—mechanical and chemical.

Mechanical weathering includes assaults on rocks by physical forces, e.g., abrasion and plucking by running water, crushing by landslides, scratching by rocks trapped at the sides of moving glaciers, and pitting from sandblasting. Rocks can also fracture from thermal stress—the expansion and contraction due to changes in ambient temperature. Biology also plays a role. Mineral grains can be plucked from rocks by the root-like structures (hyphae) of lichens. Tennessee Williams recognized this phenomenon and wrote "The violets in the mountains have broken the rocks" in his 1953 play, *Camino Real.*

One of the most effective types of mechanical weathering is frost wedging. Rocks containing fractures can get rained on and the fractures may fill with water. Because water expands by 9.2% when it freezes, water-filled fractures located away from the rock surface exert appreciable internal pressure in the rocks when temperatures dip somewhat below 0°C. This pressure acts to widen fractures and can eventually split rocks apart.

A related process (likely more important than frost wedging) is ice segregation. This mechanism begins with the accumulation of ice within pores or fractures in the rock. Liquid water is drawn toward the ice by capillary action,[1] producing ice lenses that exert high internal pressures in the rock after the additional water freezes. Salt wedging is an analogous process: (1) brine seeps into fractures, (2) the water evaporates and grains of salt precipitate, (3) additional brine is drawn toward the salt by capillary action, and (4) as the process repeats, the salt crystals grow and induce high internal pressures inside the rock. The internal rock pressures generated by frost wedging, ice segregation, and salt wedging can exceed three times the tensile strength of granite.

Chemical weathering is the set of processes that degrades rocks and minerals chemically. Greater rainfall and higher temperatures accelerate chemical weathering.

---

[1] Capillary action is the process wherein a liquid is propelled through a narrow space (commonly a thin tube) without the assistance of gravity. There is adhesion between the liquid and the solids in the tube due to differences in charge. A cross-section through the tube would show that the liquid has an M-shaped profile—higher at the sides of the tube and lower near the center. Cohesion within the liquid allows the adhering liquid molecules at the side of the tube to pull adjacent liquid molecules along with them up the tube. The narrower the tube, the stronger the adhesion with the sides and the more effective the capillary action.

A major mineralogical constituent of many igneous, metamorphic, and sedimentary rocks is feldspar. Feldspar is actually a mineral group that includes plagioclase feldspars (ranging from albite—$NaAlSi_3O_8$ to anorthite—$CaAl_2Si_2O_8$), alkali feldspars (ranging from albite—$NaAlSi_3O_8$ to orthoclase—$KAlSi_3O_8$), and celsian ($BaAl_2Si_2O_8$). We will concentrate on orthoclase, i.e., potassium feldspar (K feldspar), because it is the dominant feldspar in granite, and granite is the most abundant rock in the continental crust.

When exposed to acid, K-feldspar breaks down into kaolinite (a clay mineral): $KAlSi_3O_8 \rightarrow Al_2Si_2O_5(OH)_4$. The weak acid involved in this reaction is carbonic acid (H_2CO_3), the most abundant acid on the surface of the Earth. It forms from small amounts of atmospheric carbon dioxide dissolved in rainwater:

$$CO_2\,(gas) + H_2O\,(liquid) = H_2CO_3\,(solution).$$

Over long periods of time, carbonic acid is capable of leaching large amounts of feldspar in rocks to form abundant clay. This dissolution disrupts the interlocking mineral structures of the rocks, causing them to crumble. A simplified version of the reaction is:

$$2KAlSi_3O_8 + H_2O + H_2CO_3 \rightarrow Al_2Si_2O_5(OH)_4 + 2K^+\,(aq) + CO_3^{2-}\,(aq) + 4SiO_2\,(aq)$$

K-feldspar + water + carbonic acid → kaolinite + chemical components in aqueous solution

Calcium carbonite ($CaCO_3$), the principal constituent of limestone, can also be broken down by carbonic acid:

$$CaCO_3 + H_2CO_3 \rightarrow Ca^{2+} + 2HCO_3^-$$

to form a calcium ion and a bicarbonate ion dissolved in aqueous solution. This process leads to the chemical weathering of limestone (Fig. 4.7).

The minerals in igneous and metamorphic rocks formed at temperatures and pressures far higher than those at the Earth's surface. Under room-temperature conditions, these phases are out of chemical equilibrium; they are prone to breaking down by chemical reactions with water, atmospheric oxygen, and acids.

Olivine [$(Mg, Fe)_2SiO_4$] is a common mineral in those igneous rocks with bulk chemical compositions with relatively little silica. As an illustration, we focus on the magnesian endmember component of olivine: the mineral forsterite (Mg_2SiO_4). This phase can break down by hydrolysis, in which only a portion of the original mineral goes into solution:

$$Mg_2SiO_4 + 4H_2O = 2Mg(OH)_2 + Si(OH)_4\,(aq)$$

forsterite + water = brucite   + silicic acid in aqueous solution

Forsterite can also break down by carbonic acid in a process of carbonation:

Fig. 4.7 Weathered limestone outcrop from near Waipu Caves, New Zealand. Image from Bernard Spragg

$$Mg_2SiO_4 + 2H_2CO_3 \quad = 2MgCO_3 \; + Si(OH)_4(aq)$$

forsterite + carbonic acid = magnesite + silicic acid in aqueous solution

The ratio of mechanical weathering to chemical weathering is not constant over the Earth. Chemical weathering looms in importance in low-lying areas and in regions experiencing warm temperatures and abundant rainfall. Higher temperatures and abundant water facilitate the rate of chemical reactions. Mechanical weathering is dominant in cold mountainous regions. Frost wedging and ice segregation are important processes in cold and high-elevation areas, while gravity increases the kinetic energy of landslides.

Regardless of whether a rock undergoes mechanical weathering or chemical weathering, the rapidity of its degradation is a function of the rock's surface/volume ratio. Here's an example (Fig. 4.8): A cubic boulder, one meter (100 cm) on a side, exposed to the elements, starts off with a volume of 1 m^3, a surface area of 6 m^2, and a surface/volume ratio of 6 m^{-1}. If the boulder gets fragmented along fractures or joints by chemical reactions or by ice segregation, ideally it could form eight cobbles, each 0.5 m on a side. If we assume (unrealistically) that no material was lost, these eight cobbles collectively would have a total volume of 1 m^3, a total surface area of 12 m^2 and a surface/volume ratio of 12 m^{-1}, i.e., twice that of the original boulder.

If each of these eight cubic cobbles, 0.5 m (50 cm) on a side, is further broken down into eight smaller cobbles, 0.25 m (25 cm) on a side, the total volume would still be 1 m^3 (again assuming no loss of material), but the surface area would now be 24 m^2 and the surface/volume ratio would be 24 m^{-1}. Each halving of the linear dimension octuples the number of individual rocks (because $2 \times 2 \times 2 = 8$), doubles

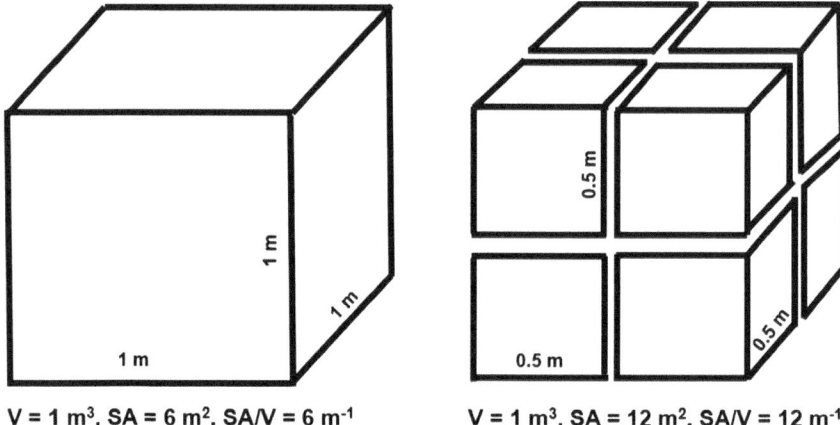

$V = 1\ m^3, SA = 6\ m^2, SA/V = 6\ m^{-1}$ $V = 1\ m^3, SA = 12\ m^2, SA/V = 12\ m^{-1}$

Fig. 4.8 A cubic boulder with a volume of 1 cubic meter could be broken down into eight cobbles with a collective surface area and surface/volume ratio twice that of the original boulder

the total surface area, and doubles the aggregate surface/volume ratio. The large amount of newly exposed surface on the disintegrating cobbles facilitates more rapid weathering (particularly chemical weathering), resulting in even smaller rocks and even faster weathering. Weathering traps rocks in this positive feedback loop until the altered, disaggregated rocks eventually are transformed into soil, a process known as pedogenesis.

And speaking of soil… A generic soil profile from a temperate climate zone is shown in Fig. 4.9. The organic layer (**O**) contains abundant plant litter. The **A** horizon (the surface layer or topsoil) consists of organic matter, living organisms, and various weathering products, including clay minerals and iron oxides. The subsoil horizon (**B** layer) contains abundant iron oxides, clay minerals, and some soluble salts; there is less organic matter than in the **A** layer. The substratum layer (**C** horizon) contains weathered, partially weathered, and unweathered bedrock. The bottom layer is the local bedrock, generally minimally weathered, upon which the soil was built. The smaller rocks in the **C** horizon have higher surface/volume ratios and break down faster than larger rocks. This causes a rough gradient of decreasing rock size from the bedrock (**R**) to the bottom of the **B** horizon.

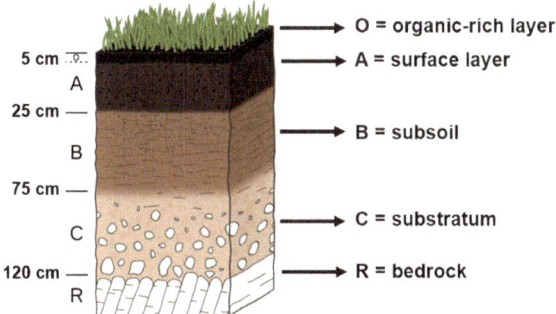

Fig. 4.9 Representative soil profile in a temperate zone showing the different soil horizons. Each layer was ultimately formed from bedrock (**R**) by mechanical and chemical weathering. Image modified from Tomáš Kebert & umimeto.org

References

Blake, S. (1981). Volcanism and the dynamics of open magma chambers. *Nature, 289*, 783–785. https://doi.org/10.1038/289783a0

BVSP—Basaltic Volcanism Study Project. (1981). *Basaltic volcanism on the terrestrial planets* (1286 pp.). Pergamon Press, New York.

Gee, J. S., & Meurer, W. P. (2002). Slow cooling of middle and lower oceanic crust inferred from multicomponent magnetizations of gabbroic rocks from the Mid-Atlantic Ridge south of the Kane fracture zone (MARK) area. *Journal of Geophysical Research: Solid Earth, 107*(B7), 2137. https://doi.org/10.1029/2000JB000062

Lofgren, G. E. (1979). Effect of nucleation on basaltic textures. *Abstract with Progrms Geological Society of America, 11*, 467–468.

Lofgren, G. E. (1983). Effect of heterogeneous nucleation on basaltic textures: A dynamic crystallization study. *Journal of Petrology, 24*, 229–255.

Mueller, R. F., & Saxena, S. K. (1977). *Chemical petrology*. New York: Springer, 394 pp.

Sen, P. N., Straley, C., Kenyon, W. E., & Whittingham, M. S. (1990). Surface-to-volume ratio, charge density, nuclear magnetic relaxation, and permeability in clay-bearing sandstones. *Geophysics, 55*, 61–69.

Usui, Y. (2013). Paleointensity estimates from oceanic gabbros: Effects of hydrothermal alteration and cooling rate. *Earth, Planets and Space, 65*, 985–996.

Waters, A. C. (1962). Basalt magma types and their tectonic associations: Pacific Northwest of the United States. *American Geophysical Union, Geophysical Monograph Series, 6*, 158–170.

Williams T. (1953). Camino Real. In M. Gussow & K. Holditch (Eds.), Williams, Tennessee. Plays 1937–1955. Library of America, 2000, New York.

Chapter 5
Geometry of Life

5.1 Metabolism

Every living organism—mushrooms and monkeys, chrysanthemums and cows, spiders and skunks, carrots and koalas—use energy, most commonly derived directly from sunlight through photosynthesis or indirectly by ingesting other organisms. Living things use this energy to perform three essential tasks: (1) powering the chemical reactions necessary to sustain life processes, (2) eliminating waste, and (3) degrading food to produce biologically important macromolecules.

The four principal macromolecules produced by organisms are:

Proteins—long chains of amino-acid residues. The amino acids are organic compounds consisting of a central carbon atom (C_α) [a.k.a. an alpha (α) carbon] bonded to H_2N (an amino group), COOH (a carboxylic group), a hydrogen atom, and a side chain (i.e., an R group or residue, typically containing various combinations of C, H, O, S, and N) (Fig. 5.1). The amino-acid residues are what remains after the amino and carboxyl groups are removed from the assembling structure when polypeptide chains or proteins join together. Proteins are involved in such essential functions as replicating DNA, forming structural frameworks for cells, and transporting molecules.

Carbohydrates—molecules consisting of carbon, hydrogen, and oxygen atoms, typically with an H/O ratio of 2/1. Common carbohydrates include sugars, starch, and cellulose—substances involved in energy storage and used as structural components in organisms.

Lipids—molecules, generally insoluble in water, which include fatty acids and waxes. They are involved in energy storage, signaling (binding to proteins), and serving as structural components of cell membranes.

Nucleic acids—molecules made up of (1) a sugar with five carbon atoms, (2) a phosphate group $(PO_4)^{3-}$, and (3) a nitrogen-bearing base. The two principal nucleic acids are RNA (ribonucleic acid) and DNA (deoxyribonucleic acid) distinguishable by whether the sugar is ribose (a simple sugar of formula $C_5H_{10}O_5$) or deoxyribose (ribose that has lost an oxygen atom, i.e., $C_5H_{10}O_4$). Nucleic acids are carriers of

A. E. Rubin, *Surface/Volume*, https://doi.org/10.1007/978-3-031-23749-2_5

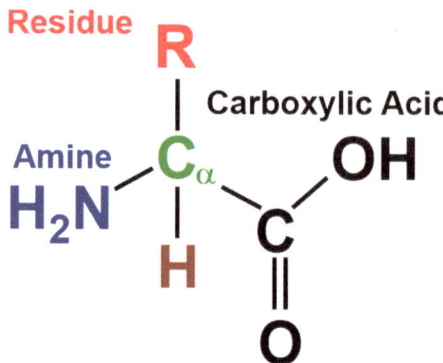

Fig. 5.1 Structure of an amino acid. Every amino acid consists of an amine group (in blue), a carboxylic acid group (black), a hydrogen atom (brown), and a residue or side chain (red R). They are all connected to a central C atom (or α-carbon) (green), designated C_α. Each amino acid has a unique residue containing C, H, and O; some also contain N and/or S

information and are essential to the operations of cell machinery. They constitute the genetic material of all living organisms, transmitting vital information about the composition of the organism from one generation to the next.

Metabolism (from the Greek *metabolē*, meaning change) is the set of all chemical reactions required to sustain life. It is the sum of the chemical activities that allow an organism to grow, produce or digest food, move, respire, fend off pests, escape predators, undergo cellular repair, and reproduce. Because there is constant change in the environment around living organisms, their metabolism must be internally regulated so the conditions within individual cells remain fairly constant. Such regulation allows organisms to receive and respond to outside information and interact successfully with their environment. In response to a signal, an enzyme (a special type of protein that acts as a catalyst) may increase or decrease its activity to regulate a specific chemical reaction in a cell without the enzyme itself being altered in the process. All metabolic activity in cells is regulated by enzymes.

The organic compound adenosine triphosphate ($C_{10}H_{16}N_5O_{13}P_3$), often referred to simply as ATP, provides the energy required for cells to function in all living organisms. In plants, ATP is produced from ADP (adenosine diphosphate—which contains one less phosphate group than ATP) during photosynthesis using the energy provided by sunlight. The photosynthesis reaction can be written as:

$$6CO_2 + 12H_2O + light - energy \rightarrow C_6H_{12}O_6 + 6O_2 + 6H_2O$$

carbon dioxide   water   glucose   oxygen   water

ATP can also be formed within mitochondria in cells during cellular respiration: (1) Aerobic respiration produces ATP, carbon dioxide (CO_2) and water from glucose (a simple sugar—$C_6H_{12}O_6$) and molecular oxygen (O_2). Most organisms use this process to obtain energy. (2) Anaerobic respiration (found in some fungi, bacteria, and archaea) requires compounds other than molecular oxygen, e.g., nitrate $(NO_3)^-$, sulfate $(SO_4)^{2-}$, fumarate $(C_4H_2O_4)^{2-}$, or elemental sulfur (S), to produce ATP. Fermentation is used by yeast and some bacteria to produce ATP. Alcoholic fermentation can be written as:

$$C_6H_{12}O_6 \rightarrow 2C_2H_5OH + 2CO_2$$
glucose   ethanol   carbon dioxide

wherein glucose produces ethanol and carbon dioxide as well as ATP. (Although slightly toxic, ethanol is the only alcohol that humans can drink safely.)

In a process called *active transport*, ATP can facilitate the movement of essential molecules (nutrients) from regions outside the cell (where they may have only low concentrations) to the zone within the cell (where the concentrations of these nutrients can be hundreds of times higher).

Diffusion is a process that behaves in the opposite manner—as a system approaches equilibrium, diffusion causes substances to move from regions of high concentration to those of low concentration. An excellent example (that anyone can explore in the kitchen) is the color-homogenization that takes place slowly after a few drops of blue vegetable coloring are released into a large bottle of water.

Although active transport may seem contrary to the Second Law of Thermo-dynamics (which states that, in an isolated system, entropy (or disorder) tends to increase), living things are *not* isolated systems—they receive energy from the outside and release waste heat to their surroundings. Metabolism is an inefficient process—about 60% of metabolic energy is converted into waste heat rather than used in the production of ATP. Heat is generated by muscle contraction and by the friction caused in the circulatory system by flowing blood. The waste heat is lost to the environment, increasing the overall disorder of the universe.

Birds and mammals are warm-blooded creatures. Each species can maintain a relatively constant body temperature, irrespective of ambient conditions (e.g., temperature, humidity, wind), if the conditions are not too extreme, by regulating its metabolism. This process, known as *thermoregulation*, allows warm-blooded organisms to maintain the body temperature at which their cellular machinery operates with optimal efficiency. These animals produce internal heat and are classified as *endothermic* (heat producing) *homeotherms* (species capable of maintaining a stable body temperature). Most commonly, their internal body temperature is higher than the surrounding air temperature. Aiding thermoregulation is insulation (commonly in the form of fur, blubber, or feathers) that diminishes the amount of heat that can escape the body.

If the air temperature is low and animals get too cold, some species will begin to shiver; these are rapid muscle contractions that use up large amounts of ATP. As a consequence, more ATP is produced inside the body, generating additional heat.

If the air temperature is appreciably higher than body temperature, some animals will sweat (e.g., humans, apes, horses, monkeys, hippos) or pant (e.g., dogs, cats, pigs, most birds) and thereby lose excess heat.

Sweating: Sweat (also known as perspiration) is a watery liquid containing trace amounts of sodium, potassium, calcium, and magnesium. It is produced by sweat glands in the skin and excreted through pores. For the liquid to evaporate and escape, it must first be transformed into a gas. This process takes energy (known as heat of vaporization or heat of evaporation). By providing and losing this heat, the body cools off. This type of evaporative cooling follows the same physical principles as swamp coolers common in homes in semi-arid regions of the world.

Panting: This process involves a significant increase in the frequency of respiration, allowing hot air to be expelled from the body. Cooler air is drawn into the body where it contacts the moist linings of the upper respiratory tract. Some of the water in these linings is transformed into gas. Evaporation and expulsion of this gas cause the body to lose heat. This is another form of evaporative cooling. Although panting uses up energy, this is compensated for by a reduction in the metabolic activity of non-respiratory muscles.

5.2 The Surface/Volume Ratios of Mammals and Birds

How does one go about measuring the surface area and volume of an animal?

Let's start with surface area. One possibility is to skin a dead animal, stretch out the skin on a flat surface, and measure it with a ruler; the skin could also be laid out on graph paper. One immediate problem is how much to stretch the skin. As pointed out by Knut Schmidt-Nielsen (1984), rabbit skins, for example, are loose and stretchable and can lead to errors of 20% or so in the determination of surface area, depending on how the skin is handled. An alternative is to wrap dead animals in paper cut to measure; the peeled-off paper can be measured with a ruler or weighed on a scale. In 1926, a couple of researchers developed a method for determining the surface area of a cow—they painted a cow with an ink roller and counted how many times the roller revolved.

German physiologist Karl Meeh proposed in (1879) that the surface area (SA) of different mammal species could be represented by a simple equation:

$$SA = k \times M^{2/3}$$

where k is a constant and M is body mass. The mass is easily determined by weighing the animal. In 1934, Francis Benedict noted the great uncertainties in k values published by different authors for different species and suggested that researchers should just use a k value of 10.0 ± 1.0. If a researcher is willing to tolerate some uncertainty, this equation works reasonably well for many animal species.

How can animal volume be measured? We can first turn to the field of physics, to the study of fluid mechanics. There is an ancient method for measuring the volume of any object by water displacement. Here is a simple example: Take a clear graduated cylinder with markings indicating the volume in milliliters. Drop a small rock in the cylinder and watch the water level go up. The difference in water level (the displacement) is equal to the volume of the rock. A non-porous rock cut into the shape of a cube, 1 cm on a side, will displace 1 ml of water, regardless of the density of the rock. This works just as well for an irregularly shaped object. A related way to measure volume is to place the rock in a vessel that is completely full of water. The overflow of water can be collected, and its volume determined; the overflow is equal to the volume of the rock.

Before we discuss another method of measuring volume by water displacement, we need to examine the equation:

$$density = mass/volume$$

This should be familiar because densities are often expressed in units of grams (a mass unit) per cubic centimeter (a volume unit), which can be written as (g/cc) or (g/cm^3) or (g cm^{-3}). Density is often represented by the Greek letter rho, ρ. We can use algebra to rearrange the terms of the equation in two ways:

$$mass = density \times volume \quad or \quad volume = mass/density$$

A third way to measure volume is to put the water-filled vessel on a scale; drop in a rock and reweigh the vessel. Note the difference in weight, divide that value by the density of the liquid (1 g/cm^3 in the case of water) and obtain the volume.

These methods result from "Archimedes' principle." Archimedes of Syracuse, the Greek physicist, astronomer, and inventor, was the first to observe that an object that sinks in a fluid will displace an amount of fluid equal to the object's volume.

Archimedes' principle can be applied to animals. Their volume can be determined by water displacement. Small animals can be placed in small vessels; large animals can be placed in large vats or tubs.

This seems like a lot of trouble and researchers love shortcuts. There is generally no need to measure the volume of an animal because mass will do just as well. Like volume, mass is a three-dimensional property. The equation above shows that mass is equal to the product of volume and density—when volume increases, mass goes up accordingly.

Let's take gold as an example for the relationship between mass and volume (Fig. 5.2). The density of gold is 19.3 g/cm^3. A cube of gold, 1 cm on a side, will weigh 19.3 g (or 0.681 oz). (As of this writing, the price of gold is about $58 per gram; our cube of gold would be worth about $1119.)

Because mass varies directly with volume (as shown for gold in Fig. 5.2), it seems clear that, whatever effects the surface/volume ratio would have on an animal, would also be true for the surface/mass ratio.

In a comprehensive monograph, Axel Hemmingsen (1960) showed a strong positive correlation (on a log–log plot) between body surface area and body mass for vertebrates (Fig. 5.3). The data points (not shown) are contained within the shaded area. Thus, body mass correlates with body volume and body surface area. This allows us to use mass alone as a stand-in for the surface/volume ratio—a second shortcut.

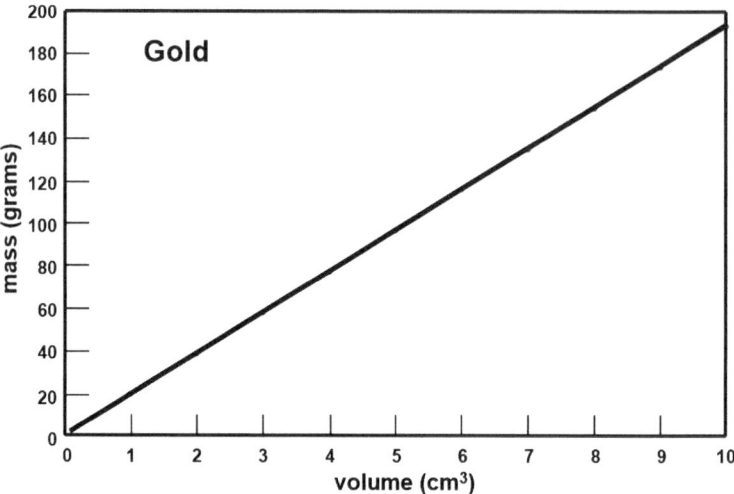

Fig. 5.2 Relationship between mass and volume of gold cubes

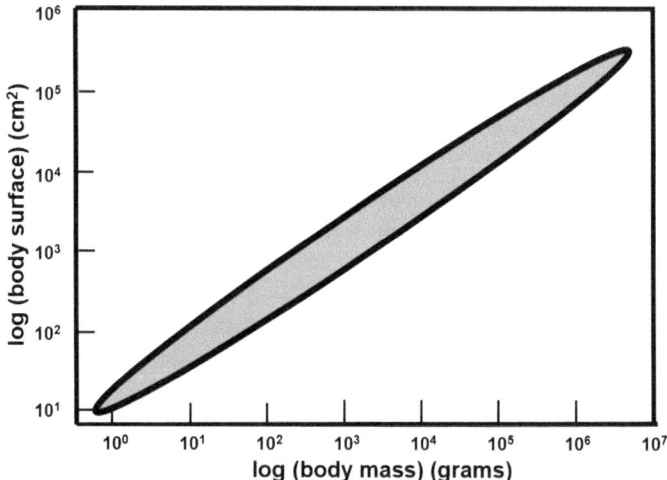

Fig. 5.3 Log–log plot showing a strong positive correlation between body surface area and body mass for vertebrates. All the data points occur within the shaded region. *Diagram* after Hemmingsen (1960)

Because mass increases as the cube of length, larger bodies require more structural support. On average, bones constitute about 8 wt.% of a mouse, 13–14 wt.% of a dog, and 16–17 wt.% of an elephant.

We need to distinguish between total metabolic activity in an animal and the specific metabolic rate per unit body mass. First, as shown in Fig. 5.4, large animals have greater total metabolic activity than small animals. This makes perfect sense. It

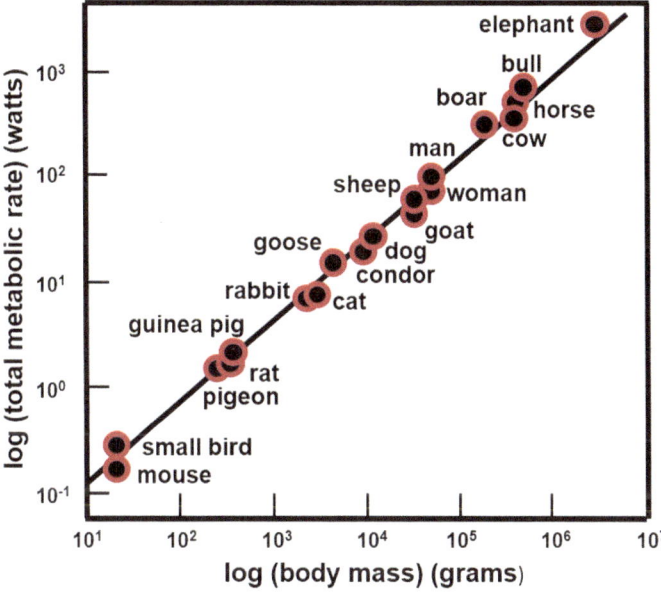

Fig. 5.4 Log–log plot of metabolic rate versus body mass for vertebrates showing a strong positive correlation

takes more total energy for an elephant to lumber across the savanna than a field mouse to scurry across a meadow. The largest male African bush elephants can weigh 10.6 tons (i.e., 10,600 kg); the smallest adult mice (the African pygmy mouse) can weigh as little as 3 g (i.e., 0.003 kg). The range in mass is more than six orders of magnitude. Francis Benedict (1934) used a log–log diagram to show the expected correlation between total metabolic rate (measured in watts) and body mass for mammals and birds (Fig. 5.4). Physiologists call this correlation the "mouse-to-elephant curve".

But what about specific metabolic rate—the metabolic rate per unit mass? One would expect that small animals (with their high surface/volume ratios and high surface/mass ratios) would lose heat faster than large animals. Heat loss would have to be compensated for by increased metabolic activity. Our prediction would be that small animals would have a higher specific metabolic rate than large animals—they should expend more energy per unit mass. As before, we can use mass alone as a stand-in for surface/volume ratio.

Physiologists have constructed an equation, based on empirical studies of numerous animals, that relates metabolic rate (P_{met}; symbolized as a P for Power, which can be expressed in kcal/day or watts/day) to body mass (M_b):

$$P_{met} = 70 \times M_b^{0.75}$$

This ratio is sometimes called Kleiber's law after the Swiss agricultural biologist, Max Kleiber. As shown by Knut Schmidt-Nielsen (1984), if we want to determine the specific metabolic rate (P*ₘₑₜ), we need to divide both sides of the equation by body mass:

$$P^*_{met} = (P_{met}/M_b) = (70 \times M_b^{0.75})/M_b = (70 \times M_b^{0.75-1.0}) = 70 \times M_b^{-0.25}$$

P^*_{met} can expressed in units of kcal per day per kilogram (kcal day^{-1} kg^{-1}). The negative exponent shows the inverse relationship between P^*_{met} (the specific metabolic rate) and body mass:

Small animals have high specific metabolic rates—they expend many kilocalories per day per kilogram of body weight.

Large animals have low specific metabolic rates—they expend relatively few kilocalories per day per kilogram of body weight.

Figure 5.5 is adapted from a schematic diagram from Knut Schmidt-Nielsen (1984). It shows that, if P^*_{met} is plotted against M_b on a log–log plot, the equation yields a regression line with a slope of −0.25.

From the equation, we can see that a small African pygmy mouse should have a specific metabolic rate of about 300 kcal kg^{-1} day^{-1}, while the largest African bush elephant should have a specific metabolic rate of about 7 kcal kg^{-1} day^{-1}. It would take more than 3½ million pygmy mice to equal the mass of one African bush elephant (Fig. 5.6); collectively and individually, pygmy mice have a specific metabolic rate more than 40 times higher than that of an African bush elephant. Gram for gram, small animals require more energy to thrive—for every single gram of sugar

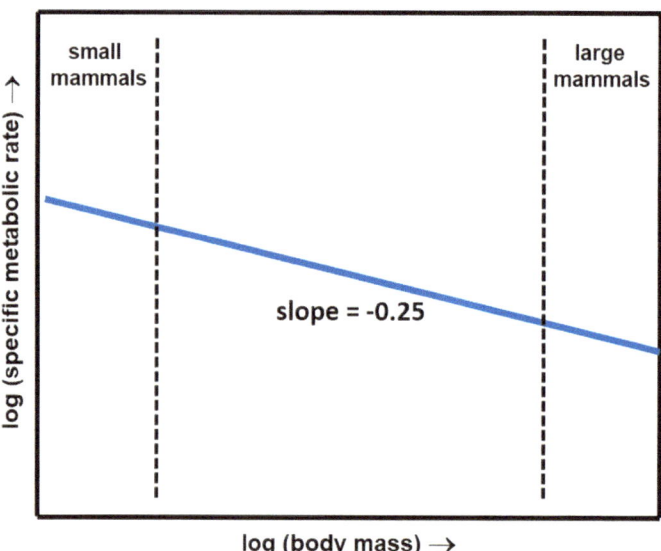

Fig. 5.5 Diagram showing that when the log of the specific metabolic rate (P*ₘₑₜ) is plotted against the log of body mass (M_b), the data form a regression line of slope −0.25. The plot is modified from one by Knut Schmidt-Nielsen (1984)

Fig. 5.6 Silhouette of elephant and mouse, showing approximate relative sizes

the average 1-kg of elephant would need to maintain its lifestyle, one kilogram of the pygmy mouse collective would need nearly 43 g of sugar. It is clear the smallest mammals have a specific metabolic rate far higher than those of the largest mammals. A similar point was raised by Bonner (2006) in his book, *Why Size Matters*: "…a huge elephant will have about 25 heartbeats per minute, while a tiny shrew's heart goes at the amazing rate of over 600 beats every minute."

However, not all mammals fall exactly along the regression line. The real world is messy. For example, marine mammals tend to have specific metabolic rates that are roughly twice as high as predicted. This is probably because cold water can remove heat from an object more than 20 times faster than room-temperature air (a result of water's high specific heat capacity). This particular problem in thermoregulation was solved in marine-mammal evolution by jacking up the specific metabolic rates of their bodies.

If mammals had a specific metabolic rate that was governed exclusively by their surface/volume ratio, the exponent in the equation (and the slope of the line in Fig. 5.5) would be −0.33 instead of −0.25. However, animals are not simple homogeneous bodies; they are not bricks. Mammals and birds have evolved strategies to retain body heat, including (in many cases) covering their outer regions with blubber, fur, or feathers. This ability to dimmish the amount of heat that is radiated away likely accounts for the difference in slope. The lower slope means these animals do not require as high a specific metabolic rate as they would otherwise need if they lost their body heat quickly via radiation into their surroundings. This means they don't require as much food to survive than they would if they were naked and fat-free.

Nevertheless, because small mammals have high specific metabolic rates, they must consume a great amount of food to power their daily activities. Because Fig. 5.4 is called the "mouse-to-elephant curve", let's start with mice. Although they are mainly herbivorous, mice are actually omnivores—they will eat seeds, grains, fruit and meat. Different species average about 15–20 meals a day, consuming roughly 10–15% of their body mass.

The average Mongolian gerbil weighs about 100 g and consumes roughly 12 g of food a day (i.e., about 12% of its body mass). Its diet consists of grass, seeds, leaves, bulbs and roots.

Let's move up the mammalian mass scale.

Domestic cats average about 4.5 kg. Although cats are finicky and their diets vary, they are principally carnivores and consume about 180 g of food a day. That is ~ 4% of their body mass.

A male chimpanzee, a mid-size mammal, may weigh about 55 kg. He will consume about 1.7 kg of food on an average day—about 3% of his body mass. Chimps prefer fruit, but will also consume a varied menu of leaves, seeds, stems, blossoms, honey, insects (ants, honeybees, and termites), birds and bird eggs, small monkeys, and warthogs.

The average human weighs about 62 kg and has a daily food intake of about 1.8 kg. That is about 3% of body mass, very similar to that of the chimpanzee (our closest living relative).

On to large mammals. A 500-kg cow, with an appreciably lower specific metabolic rate than a mouse or a cat, will consume about 2% of her body mass daily (depending on the moisture of the hay and its crude protein content).

A 6000-kg African bush elephant will consume about 150-kg of food a day (primarily grass, herbs, leaves, fruit, seeds, and occasionally tree bark)—about 2.5% of its body mass.

These observations show that larger mammals, with their lower specific metabolic rates (largely a consequence of their lower surface/volume ratios), consume less food relative to their body mass to maintain their lifestyles. The trend is strong, but the correlation is not perfect. The huge diversity of mammals and their complex interactions with their environments have led to evolutionary strategies that increase their fitness. In many cases, this allows the animals to retain more of their body heat and reduce the amount of food they must consume to live.

Because mammals and birds are both warm-blooded, it makes sense to see if there is a similar relationship between specific metabolic rate and mass for birds. Although some researchers separate different types of birds (passerine and non-passerine) in their studies of metabolic rate, we'll use a broad brush and lump all birds together. [Passerine birds have three toes pointing forward and one toe pointing backward; common passerines include finches, sparrows, robins, thrushes, cardinals, jays, wrens, turkeys, and chickadees. Non-passerines do not have this particular toe arrangement; members of this diverse club include ducks, owls, eagles, falcons, parrots, geese, swans, woodpeckers, and hummingbirds.]

The empirically derived equation that relates metabolic rate to body mass for the set of all birds is:

$$P_{met} = 86.4 \times M_b^{0.67}$$

Note the relationship has an exponent of 0.67, identical to the geometric relationship between surface area and volume. After dividing each side of the equation by M_b, we can obtain the specific metabolic rate (i.e., the metabolic rate per unit mass) for the set of all birds:

$$P^*_{met} = 86.4 \times M_b^{-0.33}$$

The exponent of −0.33 is the same as that for the regression of surface area per unit volume against volume (Fig. 2.10). Figure 5.7 is a log–log diagram showing that this relationship has a regression line with a slope of −0.33. It can be compared to the purely geometric relationship for a cube shown in Fig. 2.10. This analogy immediately suggests that, in general, the main factor governing heat loss in birds is their surface/volume ratio. If we confine ourselves to the set of all birds, it appears that their feathers are, broadly speaking, insufficient to overcome appreciable heat loss by radiation.

By analogy with mammals, it seems likely that small birds (e.g., Fig. 5.8), with their high specific metabolic rates, require large amounts of food relative to their body mass to sustain their livelihoods.

Let's start with the smallest living bird—the bee hummingbird, native to Cuba. The males have an average mass of 1.95 g and are about 5.5 cm long. Females are slightly bigger, weighing 2.6 g with a length of 6.1 cm. These birds are adapted for feeding on nectar—they have a thin pointed bill, efficient for probing flowers. Bee hummingbirds keep extremely busy: they may feed on as many as 1500 flowers a day. Their wings beat up to 200 times per second, allowing them to hover while feeding. In between flower visits, they can fly at speeds up to 48 km/hour (i.e., ~ 13 m per

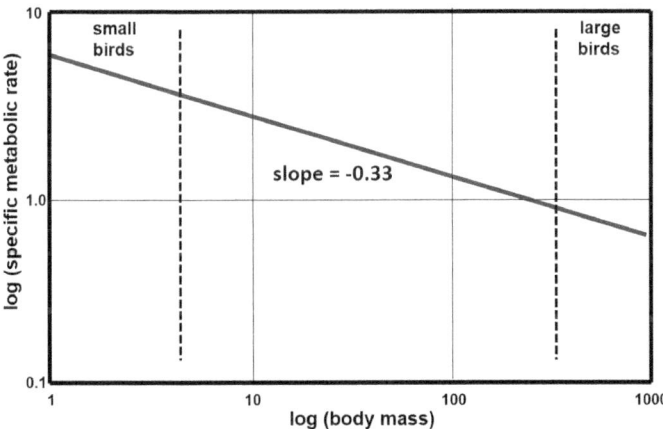

Fig. 5.7 Log–log plot showing the relationship between the specific metabolic rate and body mass for the set of all birds. The slope of −0.33 is identical to the mathematical relationship between surface area per unit volume against volume (as shown in Fig. 2.10)

Fig. 5.8 Approximate
relative sizes of a
hummingbird and a
flamingo. Combination of
vintage art images

second). Their diet also includes the occasional spider and small insect. Each day
the bee hummingbird can consume up to 50% of its body mass in food.

Macaws are intermediate-size birds. They are a variety of parrot native to South
America, Central America, and Mexico. Relative to other parrot species, macaws
have larger beaks and longer tails. Their average weight is about 925 g. They eat a
wide variety of plants including fruits, palm fruits, nuts, seeds, flowers, leaves, and
stems. On average, macaws consume about 125 g of food daily, equivalent to ~ 14%
of their body mass.

Large waterfowl deviate somewhat from the trend of decreasing daily intake of
food (relative to body mass) with increased body mass. For example, the American
white pelican weighs up to ~ 9 kg and consumes about 1.8 kg of fish a day, equivalent
to 20% of its body mass. Similarly, swans typically weigh 9–12 kg and consume about
1.8–3.2 kg of food daily, equivalent to ~ 20–25% of their body mass.

Emperor penguins, the largest living penguin species, are native to Antarctica.
They are up to 1 m in length and typically weigh between 22 and 45 kg. They
consume mostly fish but will also eat krill and squid. They take in about 2–3 kg
of food a day, equivalent to 7–9% of their body mass. In preparation for the long
winter, emperor penguins may consume about twice their daily average of food. This
is important because a male may go up to four months without eating while waiting
for the incubating egg to hatch. (After laying a single egg in late Fall, the female
departs for nine weeks of feeding.) The males rely on the food reserves they built up
during the previous summer to survive the winter.

At the high end of the avian size scale is the common ostrich, the largest living
bird. It is native to Africa. The very largest individuals can weigh up to 156 kg,
but typical Masai ostriches from East Africa weigh appreciably less: 115 kg for
males and 100 kg for females. Males range in height from 2.1 to 2.8 m, females
from 1.7 to 2.0 m. Although they cannot fly, they are fast runners, able to sustain

speeds of 55 km/hour; they can reach 70 km/hour in short bursts. No land bird is faster. Ostriches are not picky eaters—their diet consists mostly of plant matter (leaves, seeds, shrubs, roots, grass, flowers), but they will also consume locusts, lizards, rodents, and snakes. Ostriches lack teeth and are unable to tear, crush, and grind down their food before swallowing it. Instead, they swallow pebbles (known as gastroliths or stomach stones) which help grind food in the gizzard. Neglecting the mass of the non-nutritious pebbles, ostriches will consume an average of about 1.4 kg of food a day. That is equivalent to about 1.3% of their body mass. The relatively small surface/volume ratio of these large birds minimizes their heat loss; the rate of heat loss is also diminished by the year-round warm temperatures of their environment. Ostriches thus require only a modest specific metabolic rate to thrive.

Because of their high surface/volume ratios, small animals (even warm-blooded ones), can lose too much body heat in cold environments to survive. As pointed out by Haldane (1926): "In the arctic regions there are no reptiles or amphibians, and no small mammals. The smallest mammal in Spitzbergen [an island in northern Norway] is the fox. The small birds fly away in winter, while the insects die, though their eggs can survive six months or more of frost. The most successful mammals are bears, seals, and walruses" [all are large animals with low surface/volume ratios].

5.3   Muscles and Wings

Typical adult human walking (the slow gait of upright human forward motion) involves two phases (Fig. 5.9): (1) The swing phase (beginning when the foot is first lifted off the ground and ending when the heal touches the ground) and (2) the stance phase (beginning when the heal touches the ground and ending when the toes of the same foot touch the ground). Running (the fast gait) is different. It involves more full body movement including a heel strike (landing heel first), a mid-foot strike (landing on the middle of the foot), and foot rolling during landing. Arm movement is also important; during running, the arms swing toward the midline and elbows are bent at angles between about 70° and 110°.

While walking, one foot is always touching the ground; while running, both feet are off the ground simultaneously most of the time. At a high fixed speed, running expends less energy than fast walking. [The sport of fast walking, known as racewalking, was cinematized in the 1966 film, *Walk, Don't Run*. Set at the 1964 Summer Olympics in Tokyo, it was Cary Grant's final film role. But because many find the sport even more boring than the movie, the 50-km racewalking event was dropped from the Summer Olympic schedule after 2021.]

Most birds and many bats also exhibit two distinct gaits while flying—slow and fast. This gait is the wing-beat pattern employed when flying at different speeds. At slow flight speeds, the slow gait requires less energy than the fast gait; at fast flight speeds, the fast gait is more energy efficient than the slow gait.

Flying creates turbulence in the air in the flyer's wake; the type of gait is defined by the kinds of vortices the flying generates. The slow gait produces ring-shaped vortices

Fig. 5.9 The human fast
gait and slow gait involve
positioning the limbs and
using the muscles in
different ways

RUNNING   **WALKING**

(resembling smoke rings) after each downstroke of the wings. The fast gait produces
a continuous string of oscillating cylindrical patches of air streaming backwards from
each wing tip. If a slow-flying bird is moving its wings faster than its forward speed, it
is likely using the slow gait. An extreme example of this is a hovering hummingbird.
In contrast, if a bird is flying forward at a speed greatly exceeding its rate of wing
flapping, it is probably using the fast gait.

Lift is produced during the wing downstroke; lift is perpendicular to the direction
of airflow. Drag is parallel to the airflow. The resultant aerodynamic force (which is
tempting to call the RAF) is the vector sum of the lift and drag (Fig. 5.10); during the
downstroke, the resultant aerodynamic force is tilted forward and produces thrust.
For the bird to fly forward, thrust must exceed drag.

An important physical feature of wings is their aspect ratio. This parameter is
defined as the square of the wingspan (i.e., the square of the width) divided by the
wing area. Long, narrow wings have high aspect ratios; short, wide wings have low
aspect ratios.

In general, large birds fly faster than small birds. As discussed by Alexander
(2002), small birds (including most passerines) may be restricted to using the slow

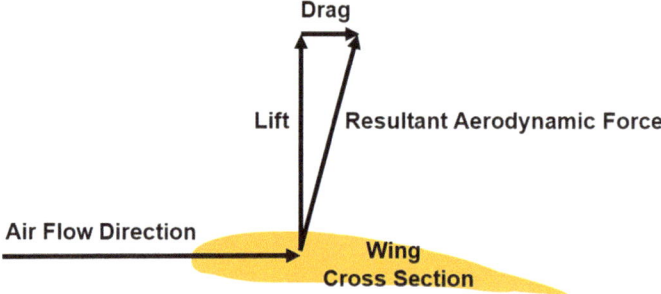

Fig. 5.10 Aerodynamic relationships. The resultant aerodynamic force is the vector sum of lift and
drag

Fig. 5.11 Wandering albatross (*Diomedea exulans*) near Tasmania. *Image* from J.J. Harrison

gait due to their slow flight speeds, low lift/drag ratios, and low wing aspect ratios. They tend to have high wing-beat frequencies. Hummingbirds beat their wings 30–90 times per second. Large birds (including gulls and large falcons) have wings with high aspect ratios; they spend most of their flight time using the fast gait. Gulls beat their wings in fast flight 10–12 times per second; when not soaring, the Andean condor beats its wings only 1–3 times per second. Albatrosses (Fig. 5.11) have the highest wing aspect ratio among flying birds; they probably use the fast gait exclusively.

Why do very large birds rarely fly slowly? The answer is that flight speed is dictated by the bird's surface/volume ratio. If a large bird is twice as long as a small bird, the large bird has four times more surface area and eight times more volume. Because volume and mass are correlated three-dimensional properties, the large bird is also eight times heavier than the small bird. As a bird gets larger, its mass increases much faster than its wing area. Thus, a large bird must fly much faster to offset its increased mass. To fly slowly or hover, a bird must flap its wings at a very high rate to compensate for slow flight. Large birds are generally too heavy to accomplish this.

The flying ability of birds is rooted in their muscles (Fig. 5.12). The largest muscles in birds are the pectorals which control the wings. Beneath these muscles are the supracoracoideus muscles which raise the wings between beats. These two muscle groups together constitute 25–40% of total bird-body weight.

Effective muscle strength is also governed by the surface/volume ratio. Specifically, raw muscle strength is a function of the muscle's cross-sectional area. Whereas the cross-sectional area increases as the square of length, muscle mass increases as the cube. Here is an example: A muscle in a small bird is 2 cm long; the comparable muscle in a large bird is 8 cm long. The larger bird has a muscle with 16 times the cross-sectional area of the small bird and can generate 16 times more force. However, the muscle of the larger bird weighs 64 times as much as that of the small bird. The larger bird's muscle has a surface/mass ratio (and surface/volume ratio) only 25% as great as that of the small bird. The muscle of a large bird produces less force per

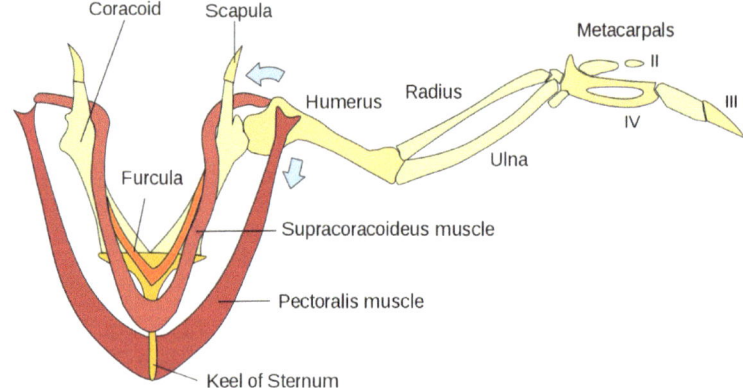

Fig. 5.12 Musculature and skeletal structure of a bird's wing. *Image* from L. Shyamal

kilogram of bird than the muscle of a small bird. Large birds have proportionately less muscle area than small birds; large birds thus lack the muscular strength to flap their wings fast enough to use the slow gait for sustained periods.

Some large birds such as hawks and falcons spend much of their airtime soaring, maintaining flight without flapping their wings (Fig. 5.13). They use rising warm air currents to keep them aloft. Many of these birds can lock their wings by using a specialized tendon. But this particular physiological specialization, that is a great aid to soaring, hampers the use of the slow gait (Alexander, 2002).

In contrast, small birds (as well as small bats and large insects) have low lift/drag ratios; they fly too slowly to use the fast gait. But they have muscles with high (cross-sectional-area)/mass ratios, providing them with sufficient power for slow flight and hovering.

Medium-size birds and bats can probably use both the slow gait and the fast gait, depending on immediate need. However, unlike the small, but mighty hummingbird (Fig. 5.14), medium-size birds are too big to hover for extended periods.

Fig. 5.13 Andean condor soaring on rising air currents over the Colca Valley in Peru. *Image* by Colegota

Fig. 5.14 Purple-throated Carib hummingbird (*Eulampis jugularis*) in Dominica, hovering while feeding. *Image* by Charles J. Sharp

5.4 Formidable Formicidae or the Mighty Ant

There are roughly 22,000 ant species marching across the Earth; they are native to every continent except Antarctica and inhabit all but a few inaccessible islands. In most land-based ecosystems, ants constitute a significant fraction of the animal biomass. One recent study (Schultheiss et al., 2022) estimated there are nearly 20×10^{15} (i.e., 20 quadrillion) individual ants, equivalent to 12 megatons of dry carbon, about one-fifth of the total human biomass. Ants belong to the family Formicidae and join wasps, bees, and sawflies in the order Hymenoptera.

Extant ants come in a variety of sizes, ranging from 0.75 mm to 5.2 cm. Ant bodies (Fig. 5.15) are divided into three principal segments (with some small variations in individual parts among different species): the head (sporting two large compound eyes, three small simple eyes (ocelli), two elbowed antennae, and two mandibles), the thorax or mesosoma (to which are attached six legs and, in reproductive individuals (queens and males), four wings), and the abdomen or metasoma (which houses internal organs used for reproduction, respiration and excretion). The petiole is a narrow waist located between the thorax and the abdomen; the petiole is fused to the first abdominal segment. The gaster is the remaining, bulbous portion of the abdomen.

Ants have an exoskeleton, a light-weight, waterproof, structure made of chitin, $(C_8H_{13}O_5N)_n$—a long-chain polymer derived from glucose $(C_6H_{12}O_6)$, to which the muscles are attached. Depending on the species, the thickness of the exoskeleton can range from about 1 μm to 110 μm; larger species tend to have thicker exoskeletons.

Ants do not have lungs. (No insects do.) Air enters their bodies through small respiratory ducts called spiracles; ants have nine or ten pairs on the sides of their bodies, one pair per body segment. These openings lead to a network of channels called tracheae that carry molecular oxygen (O_2) throughout the body, equalizing the

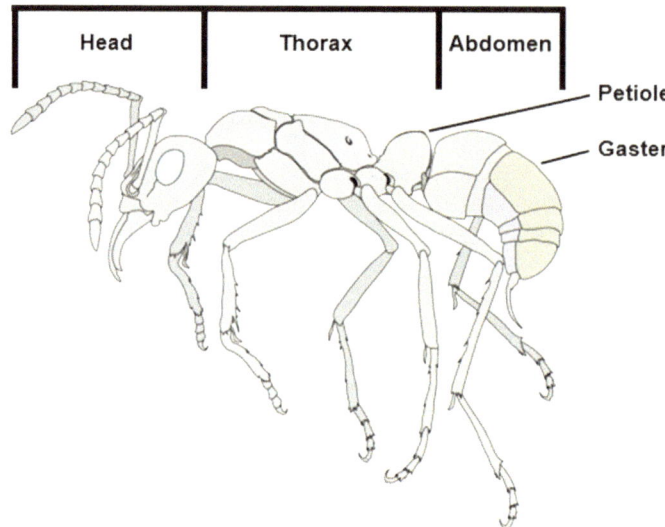

Fig. 5.15 Generalized anatomy of an ant. *Image* modified from one by LadyofHats and Sophivorus

pressure of the gas. The tracheae narrow with distance from the spiracles, with each trachea terminating at a cell membrane. Oxygen is not pumped through the body; it diffuses through the tracheae. The high surface/volume ratios of the fine tracheae allow molecular oxygen to diffuse rapidly to all the ant's cells; it also allows carbon dioxide (CO_2) to diffuse from the cells and out of the body through the spiracles.

Ants are impressively strong. Leaf-cutter ants (Fig. 5.16) can carry snipped leaf fragments up to 20 times their body weight. Wojtusiak et al. (1995) reported a weaver ant (*Oecophylla longinoda*) supporting a dead 7-g bird that weighed 1200 times as much as the ant. Mechanical testing of the Allegheny mound ant (*Formica exsectoides*) showed it could withstand a load equivalent to ~ 5000 times the ant's weight (Nguyen et al., 2014).

Why are ants so strong? As discussed in the previous section, muscle strength (a function of cross-sectional area) increases as the square of length, while muscle mass (a three-dimensional property) increases as the cube of length. Thus, large animals have high muscle-mass/strength ratios; small animals, like ants, have high muscle-strength/mass ratios. Furthermore, the light-weight exoskeleton of ants (and other insects) results in less body mass than if they instead had small bones inside them. Because relatively little energy is required for ants to move their small, unencumbered bodies, more muscle power is available to tote heavy loads.

Fig. 5.16 A worker
leaf-cutter ant carrying a leaf
back to the colony. *Image* by
Scott Bauer, U.S.
Department of Agriculture

5.5   Biological Rules of Thumb

5.5.1   Bergmann's Rule

Because small animals tend to have higher surface/volume ratios than large animals, they tend to dissipate their body heat more quickly (i.e., they radiate away more body heat per unit of mass). One could conjecture that the proportion of small animals to large animals would be greater in warm climates where there would be less chance for small animals to freeze to death. In cold climates, one would expect larger animals to dominate. Generally, larger animals are heavier than small animals because volume and mass both increase as the cube of the length (e.g., Fig. 5.2). The supposition would be that there is a rough correlation between animal body size (or animal weight) and latitude. This postulate is known as "Bergmann's Rule", named after German biologist Carl Bergmann, who independently discovered this relationship in 1847. It proposes that, within a single wide-spread species, the populations residing in cold climates have larger average sizes and greater mean masses than those residing in warm climates.

Because the real world is messy, this "rule" is not iron clad, but seems to apply (roughly) to many birds and mammals. For example, the largest living species of penguin is the Emperor penguin (*Aptenodytes forsteri*) (100 cm tall; 22–45 kg); it resides in Antarctica at high latitudes: 66° to 77°S. Smaller penguins reside at lower latitudes. The Southern Rockhopper penguin (*Eudyptes chrysocome chrysocome*) (45–58 cm; 2.0–3.4 kg) is native to the Falkland Islands (51° 42′S); the Galápagos penguin (*Spheniscus mendiculus*) is of similar size (49–50 cm; 2.5–4.5 kg) and lives near the equator (0° 22′S). However, the smallest penguin species, the Little penguin (*Eudyptula minor*) (33 cm; 1.5 kg) resides mainly in Tasmania (42°S), i.e., at much higher latitudes than the Southern Rockhopper or the Galápagos penguins. Although the correlation between penguin body size and latitude is not perfect, it is clear the largest living pinguin species (the Emperor penguin) is restricted to high latitudes.

The Siberian tiger (*Panthera tigris tigris*) lives at high latitudes (roughly 60°N) where the average January temperature is about −25 °C. Males weigh between 180 and 306 kg, females between 100 and 167 kg. In contrast, the Sumatran tiger (*Panthera tigris sondaica*) lives near the equator. November is the coolest month with an average temperature of 31 °C; the typical minimum temperature over the course of a year is 23 °C. Male Sumatran tigers range in weight from 100 to 140 kg, females from 75 to 110 kg. Although the Siberian and Sumatran tigers are members of the same species, they evolved different body sizes in response to the local climate in accordance with Bergmann's Rule of thumb.

The cougar (a.k.a. puma or mountain lion; *Puma concolor*), a large cat native to North and South America, also seems to follow Bergmann's Rule. There is a general trend of increasing mean body size at higher latitudes. Small and very small cougars are restricted to low latitudes; large and very large cougars are restricted mainly to high latitudes (north and south). Medium-size cougars are adaptable to varied climatic conditions; they prowl over a wide latitudinal range (Fig. 5.17).

Because Bergmann's Rule is no more than a rule of thumb, there are numerous exceptions, noted by numerous critics. Canadian biologist Valerius Geist (1987) pointed out that animal body size is proportional to food availability. If food is

Fig. 5.17 Distribution of the cougar in the Americas, ranked by mean body size. VL = very large; L = large; M = medium; S = small; VS = very small. After Newman (1953)

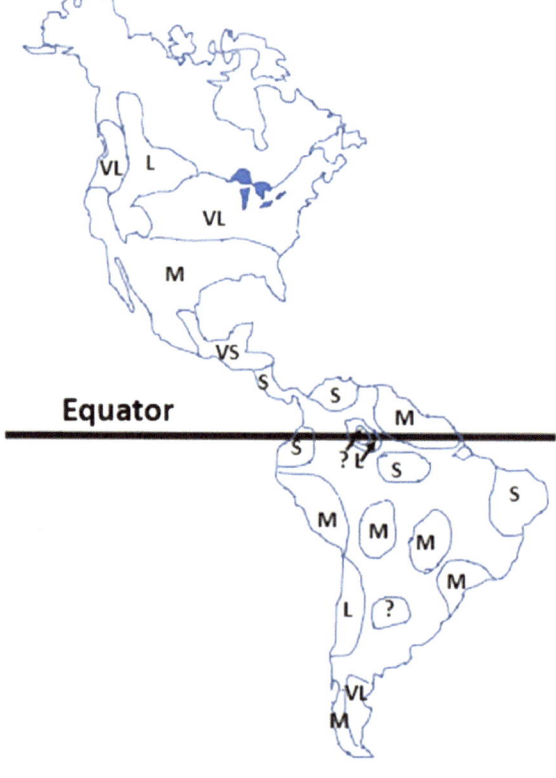

scarce in an environment, there would be evolutionary pressure for animals to become smaller; this would result in fewer resources being required for each individual.

5.5.2 Allen's Rule

Formulated by American zoologist Joel Allen in 1877, Allen's Rule holds that the surface/volume ratio or surface/mass ratio of the body parts of endothermic (warm-blooded) animals vary with the mean annual temperature of their environment. Thus, animals residing in cold climates (generally at high latitudes) would tend to have thicker, heavier limbs than animals of similar size residing in warm climates. Thick limbs preserve heat more readily (due to their low surface/volume ratios) than thin limbs. Evidence suggests that cartilage in vertebrate species grows faster at higher temperatures; this could account for animals in warmer climates having longer, thinner limbs.

A prime example of an animal following Allen's Rule is the polar bear (*Ursus maritimus*); it is the largest living species of bear and the largest living land carnivore. Adult males generally weigh 350–700 kg; the heaviest polar bear ever recorded weighed ~ 1000 kg. Polar bears are confined mainly to the Arctic Circle. Compared to brown bears (their closest living relative, thriving at lower latitudes), polar bears have thicker limbs.

Terns and gulls (related seabirds of the family Laridae) have legs that are only partially covered with feathers. Nudds and Oswald (2007) found that the length of the portion of the leg in these birds that is not covered with feathers correlates with the difference between bird body temperature and the minimum ambient temperature during the breeding season. (Exposed leg length does not correlate with latitude.) Longer bare legs result in a greater surface area of exposed skin (Fig. 5.18), facilitating heat loss. Minimization of heat loss during the coldest part of the breeding season appears to be the principal selection pressure affecting the length of exposed leg in the seabirds, in compliance with Allen's Rule.

5.5.3 Hesse's Rule

A corollary to Bergmann's Rule and Allen's Rule is Hesse's Rule, also known as the heart-weight rule. It was proposed in 1937 by German zoologist Richard Hesse (e.g., Baum, 1997). He found that animals living in colder climates tend to have a larger heart relative to their body weight, i.e., a higher (heart-mass)/(body-mass) ratio, than animals of the same species inhabiting warm climates. Because it takes extra metabolic work to maintain a constant body temperature in cold environments, individuals from cold regions develop larger and more massive hearts.

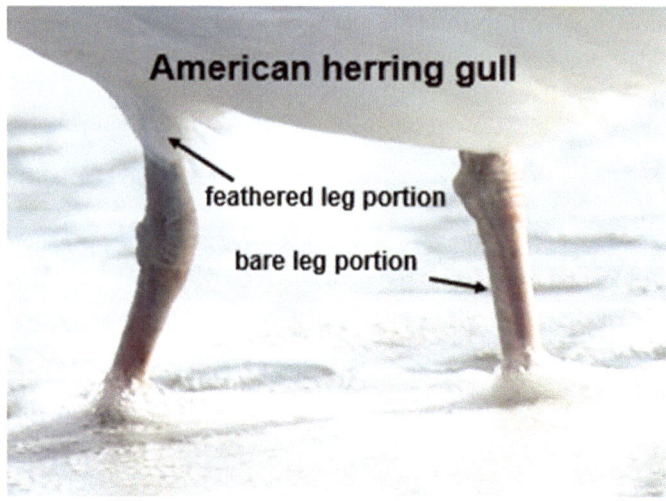

Fig. 5.18 Feathered and bare leg portions of the American herring gull

5.6   Gigantothermy

Cold-blooded animals (ectotherms) include reptiles, amphibians, fish, and inverte-brates. They require sources of heat outside their bodies (such as sunlight or solar-heated rock surfaces) to regulate their internal body temperature. Small ectotherms tend to have high surface/volume ratios and lose their heat quickly to their surround-ings. Large ectotherms tend to have low surface/volume ratios and are much better able to retain their body heat. If two cold-blooded animals are of similar shape, the larger one has a smaller proportion of its body exposed to the environment. This phenomenon is known as gigantothermy; it keeps the internal body temperatures of large ectotherms similar to those of warm-blooded animals (endotherms).

Gigantotherms include modern giant tortoises (which can reach 1.3 m in length and exceed 400 kg; e.g., Fig. 5.19) as well as extinct marine reptiles such as ichthyosaurs (one species of which, *Shonisaurus sikanniensis*, may have exceeded 20 m) and mosasaurs (one species of which, *Mosasaurus hoffmannii*, may have reached 17 m).

An advantage enjoyed by gigantotherms relative to endotherms of similar size is that the gigantotherms have a slower metabolic rate. Because it takes them longer to digest their food, they need food less frequently. This is an asset in resource-poor environments.

Fig. 5.19   An Aldabra giant tortoise (*Aldabrachelys gigantea*), native to the islands of the Aldabra Atoll in the Seychelles. This particular tortoise resides at the Beauval Zoo in France. *Image* from Yotcmdr

5.7   The Impossibility of King Kong (But Not Dinosaurs)

Synopsis: In the original 1933 film, the giant fictional gorilla-like monster, King Kong, lived on the tropical "Skull Island" in the Indian Ocean (although by Bergmann's Rule, one would expect to find him in the cold at high latitudes). An American film crew managed to capture Kong and bring him to New York City for exhibition. He became enraged, broke his chains, and escaped. Kong climbed the Empire State Building to protect a beautiful actress, with whom he was infatuated and had declined to eat earlier on the island. (Because the metabolic rate of this giant ape would be so slow, he wouldn't need to eat often.) While clinging to the Empire State Building, Kong was fired at by buzzing biplanes and fell to his death. After viewing this fatal spectacle, the filmmaker remarked "… it wasn't the airplanes. It was Beauty killed the Beast".

How big was King Kong? According to Wikipedia, his height varied in the 1933 movie depending on the setting—from 5.5 m on Skull Island to 7.3 m in New York; in some scenes he appears to exceed 18 m. In the 2017 film, *Kong: Skull Island*, Kong is 31.7 m tall; in the 2021 film, *Godzilla versus Kong*, he is a whopping 102.7 m.

How big are gorillas? The largest male gorillas are about 1.8 m tall and weigh 270 kg. The 2021 Kong is 57 times taller than the biggest gorilla. Because mass scales with the cube of length, this version of Kong is 57^3 or 185,193 times heavier than a large gorilla. That would be 50 million kilograms—twice as heavy as the Statue of Liberty attached to its base and one-fifth as heavy as the Empire State Building Kong had climbed.

Gorilla bones are much thicker and denser than human bones (due to a higher density of spongy bone), but they would certainly break under a weight of 40–50 times the full body weight of a large gorilla. The 2021 King Kong is more than 185,000 times heavier than a gorilla; he would shatter his bones with every step. Even the New York City version of the 1933 Kong would be 67 times heavier than a large gorilla and would also be unable to walk.

Because muscle strength increases as the square of length and muscle mass as the cube, Kong would have an enormous muscle-mass/strength ratio. The muscles of the larger Kong versions would be so heavy, the mighty giant wouldn't be able to lift anything at all, not even a svelte actress. Although it has been reported that a large gorilla could lift 27 times its body weight, Kong couldn't do that. Kongs are impossible.

But dinosaurs are *not* impossible—they roamed the Earth for more than 180 million years. During the Cretaceous Period (145–66 million years ago), sauropods (large, long-necked dinosaurs with small heads, long tails, and thick, trunk-like legs) grew to enormous sizes. At different stages of the Late Cretaceous, *Dreadnaughtus* may have reached 59,000 kg and *Argentinosaurus* 90,000–100,000 kg. How could their bones support them?

There are two types of bone: compact bone (a.k.a. cortical bone) and spongy bone (a.k.a. trabecular bone); the latter is much less dense and more flexible than compact bone. Many of the spaces in spongy bone are filled with marrow and blood vessels.

Aguirre et al. (2020) studied dinosaur and mammal bones and found: (1) Although trabecular bone thickness increases in mammals with body size (to diminish the strain on individual bones), it remained fairly constant in dinosaurs. (2) Larger dinosaurs instead had a higher density of spongy bone (i.e., decreased trabecular spacing). The pores in spongy bone were smaller and bone tissue was closer together than those in small dinosaurs. (3) Large dinosaurs had increased trabecular connectivity density, a characteristic that protects cortical bone by diffusing stress through the trabecular network. Well-developed trabecular networks result in greater bone strength. The skeletons of some giant dinosaurs were also lightened by networks of air sacs that extended into their bones from their throats and lungs. The bones of large dinosaurs were thus structurally different from those of the large mammals that evolved millions of years later. Even so, the largest extant land mammal—the African bush elephant (which weighs up to about 10,000 kg)—is still about an order of magnitude less massive than the very largest dinosaurs from the Late Cretaceous.

Although the non-avian dinosaurs were wiped out 66 million years ago, a few tyrannosaurs appear to have survived in the cinematic universe, battling King Kong in the 1933 and 2005 movie versions. In each case, impossible Kong was victorious.

5.8   Thermoregulation of Insects

Although insects are not endothermic, flying insects produce appreciable internal heat during flight. The second law of thermodynamics mandates that muscle power used for flight produces waste heat due to mechanical inefficiencies; in the case of insects (and birds) the total mechanical inefficiency is ~ 10–20%. Studies of insects have shown that higher flight speeds cause increases in the temperature of the thorax. Some large insects (e.g., bumblebees and many species of moth), with their comparatively low surface/volume ratios (that is, relative to other insects), can become essentially endothermic during flight—waste heat is built up in the thorax faster than it can be radiated away. Flight muscles are highly metabolically active tissues, and the temperature of these muscles can increase by 20–30 °C above the ambient temperature (Heinrich, 1974). When not flying or basking in the Sun, these same insects maintain their typical ectothermic metabolic rate.

If the ambient temperatures are high, large flying insects could be in danger of overheating. To avoid this problem, sphinx moths (Fig. 5.20) can transfer warm extra-cellular fluid (hemolymph) from the thorax to the (only lightly insulated) abdomen (which serves both as a heat sink and a site from which heat can be radiated away). These moths are thus able to maintain a steady thoracic temperature during flight, irrespective of non-extreme ambient temperatures.

In contrast, the very high surface/volume ratios of small insects (e.g., midges and fruit flies) prevent them from becoming effectively endothermic during flight. The waste heat produced by their flight muscles is radiated away too quickly to increase their thoracic temperature more than about 1 °C.

If a flying insect lands in a shaded area, it can radiate away its heat quickly due to its high surface/volume ratio. Because a high thoracic temperature must be achieved in large insects before flight can begin, they solve this problem by vibrating their flight muscles in a manner that resembles shivering. Nearly as much waste heat is produced by shivering as by flying. Whereas stationary bumblebees perched on

Fig. 5.20 Sphinx moth. *Image* from Lisafern

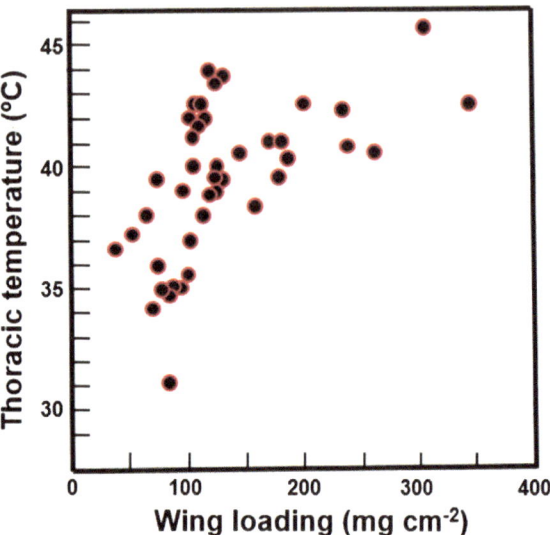

Fig. 5.21 Thoracic temperatures during free flight for 25 species of sphinx moth relative to wing loading (i.e., wing-mass/wing-area). Diagram modified from Heinrich (1974)

flowers can maintain high body temperatures on cold days by shivering, small flies (with their higher surface/volume ratios) cannot. They lose heat too rapidly.

Wing loading is defined as the ratio of the total weight of the wing to the wing area; it is measured in mass per unit wing area (e.g., mg cm^{-2}). Because weight (or mass) is a three-dimensional property like volume, wing loading is directly related to the volume/(surface-area) ratio of the wing (i.e., the inverse of the surface/volume ratio). There is a positive correlation between the thoracic temperature in members of different species of sphinx moth during free flight and wing loading (Fig. 5.21). In other words, those sphinx moths that have greater wing loading achieve higher thoracic temperatures while flying. This also indicates that sphinx moths sporting wings with high surface/mass ratios (or high surface/volume ratios) generally fail to raise their thoracic temperatures appreciably because too much heat is radiated away.

5.9   The Risk of Animal Dehydration

As pointed out by Howard et al. (2020), the "greatest physiological threat to terrestrial life is dehydration." Animals can lose water through elimination of waste, lactation, ejaculation, sweating, respiration, tear production (Fig. 5.22), and evaporation from the animal's surface. This surface, known as the integument, is the physical barrier between the interior of the organism and the external environment. Some terrestrial animals have surfaces that are more resistant to water loss than other animals.

Fig. 5.22 A uniquely human mode of water loss. The pool of tears shed by a distraught Alice when she was a giant. *Drawing* by John Tenniel from Lewis Carroll (1865): *Alice's Adventures in Wonderland*

Water loss in terrestrial animals is affected by air temperature (the higher the temperature, the greater the loss) and relative humidity (the lower the relative humidity, the greater the loss). For most animals, the principal avenue for water loss is cutaneous—i.e., through the skin. The principal driving force for water loss from an animal is the difference in vapor pressure between water within the animal and ambient water in the environment.

Cutaneous water loss is strongly affected by an animal's surface/volume ratio. All else being equal, small animals, with their high surface/volume ratios, would lose water faster than larger animals.

Many small animals (Fig. 5.23) have adopted survival strategies to diminish water loss. For example, animals in hot arid regions can be most active at night or during cooler hours of the day. Many of these animals crawl into burrows during the heat of the day. Among these heat-averse animals are the sand cat (*Felis margarita*), 40–50 cm long; the desert hedgehog, (*Paraechinus aethiopicus*), 14–28 cm; and various jerboa (e.g., the thick-tailed pygmy jerboa, *Salpingotus crassicauda*), 4.5–6 cm. The elf owl (*icrathene whitneyi*), 12.5–14.5 cm, hunts at night and resides during daylight hours in holes it finds in saguaro cacti or trees.

African bullfrog (*Pyxicephalus adspersus*) males grow up to 24.5 cm in length; females are smaller. When the weather is hot and dry, these bullfrogs dig burrows and enter a prolonged (months-long) period of dormancy known as estivation. They shed

Fig. 5.23 Small desert animals. **a** African bullfrog, image from Wikimedia commons. **b** Sand cat. **c** Elf owl, image from Dominic Sherony. **d** Desert hedgehog, image from Chen Ein-Dor

their skin and use it as a protective layer to diminish water loss. These sloughed-off skins can even absorb water from the bullfrog's bladders.

Howard et al. (2020) created a lab exercise for college students to test whether the surface/volume ratio of an animal significantly affects water loss. Because some students might find the use of live animals morally objectionable, students prepared model frogs of three different sizes made from gelatin. These gelatin frogs were subjected to wind currents and high temperatures in various experiments to promote water loss. The frogs were weighed before and after each experiment to determine how much water escaped. Figure 5.24 shows that there was a strong inverse correlation between the size of the gelatin frogs and the amount of water loss—the smallest set of frogs (i.e., the group with the highest surface/volume ratio) lost the most water. This simple experiment demonstrates that the surface/volume ratio has a dramatic effect on organisms. It puts evolutionary pressure on animals, particularly small ones, to change their physiology and/or change their behavior to minimize water loss.

Fig. 5.24 The amount of water loss (units of mg water lost per gram of original wet gelatin frog mass per minute) for model frogs of different lengths. Frogs were subjected to room temperatures for at least 20 min; in some experiments a heat lamp was used to promote water loss. *Diagram* adapted from Howard et al. (2020)

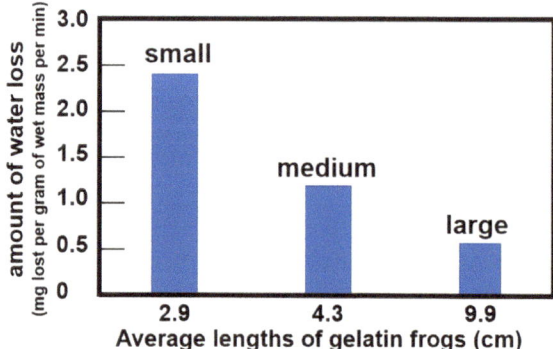

5.10 Krill

Crustaceans comprise a large group of arthropods that includes crabs, lobsters, shrimp, prawns, barnacles, and krill. It is the krill that is of interest here. They are found in all the oceans of the world; they are the main food source for baleen whales. [Humans sometimes also eat krill (e.g., the Japanese dish *okiami*); although rather salty, they are rich in protein, Vitamin A, and omega-3 fatty acids.] Most species of krill are bioluminescent, a characteristic presumably used for communicating essential information about swarming and breeding. Typical adult krill lengths reach 1–2 cm; adults of the largest krill species can reach 15 cm. All krill have an exoskeleton composed mainly of chitin (a long-chain polymer, $(C_8H_{13}O_5N)_n$). Their bodies (Fig. 5.25) consist of three main parts—a cephalothorax (the head and thorax fused together), the pleon or abdomen (which contains five pairs of swimming legs, earning them the label "decapods"), and the tail fan (containing posterior appendages and the anus). Krill heads have compound eyes, two antennas, and external gills. [The round eyes of the Arctic and subarctic krill species, *Thysanoessa inermis*, are adapted to detect very slight differences in the minute amounts of sunlight that penetrate the water during long Arctic nights.] Protruding from the abdomen of krill are several pairs of thoracic legs (the exact number depends on the species), including legs for feeding and legs for grooming.

Most krill are filter feeders. They use a "feeding basket", comprised of thoracic legs, to collect phytoplankton (and some zooplankton) from the water. [Phytoplankton are photosynthetic microorganisms, typically a few tens of micrometers in size, which serve as the krill's principal food source. Zooplankton are animals; they do not produce their own food. In fact, they eat phytoplankton.] Each krill appendage sports elongated sets of stiff bristles called setae (Fig. 5.26). The thick individual bristles in the primary set (first-degree setae) anchor two sets of secondary bristles (second-degree setae) that form V-shaped micronets. Extending from each of the secondary bristles are tiny third-degree setae. This arrangement forms a mesh finer than 1 μm.

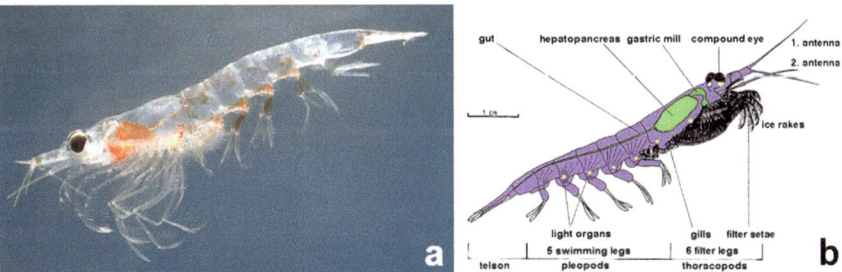

Fig. 5.25 a Northern krill (*Meganyctiphanes norvegica*). Image courtesy of Øystein Paulsen. **b** Krill anatomy as represented by *Euphausia superba. Diagram* courtesy of Uwe Kils

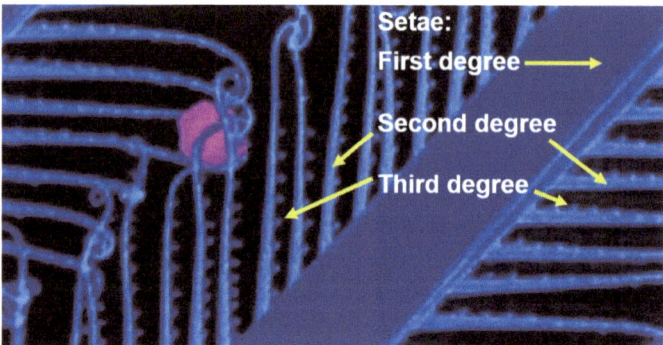

Fig. 5.26 Detail of scanning electron microscope (SEM) image of the feeding basket of an Antarctic krill. The small red polygonal object (1 μm in diameter) is probably a bacterium. *Image* courtesy of Uwe Kils

There are different feeding behaviors exhibited by krill. One behavior involves a pumping action accomplished by periodically opening and closing the feeding basket by swinging the thoracic appendages to the side and front and pulling them back. Where plankton are scarce, the feeding basket can remain open to collect more microorganisms as the krill swims forward for half a meter or so.

It is the enormous surface area and high surface/volume ratio of the combined set of first-, second-, and third-degree setae that permit krill to sift the water and sweep up sufficient quantities of plankton.

The longevity of krill species correlates with latitude. For example, *Euphausia superba* live at high latitudes and last more than six years. *Euphausia pacifica* live at mid latitudes and survive about two years. *Nyctiphanes simplex* lives in the tropics and subtropics (near Baja, California), but survives only six to eight months.

5.11   Sponges

Sponges are porous multicellular animals whose bodies have evolved to maximize water flow through their central cavity (Fig. 5.27). This flow facilitates the uptake of food and oxygen from the bottom of the sponge and the expulsion of wastes (including carbon dioxide and ammonia) at the top (the osculum). There are about 5000 species of sponge inhabiting all the world's oceans, from high latitudes to the equator. They tend to live in clear, calm waters, minimizing the chances that suspended sediments would block their pores. Once adult sponges attach themselves to surfaces (e.g., rocks, reefs), they remain there.

The bodies of most sponges consist mainly of non-living gelatinous material called mesohyl that serves as an exoskeleton. In many cases, mesohyl is supported by fibrous skeletal structures composed of spongin (secreted collagen proteins) and/or spicules (tiny sharp spikes) composed of calcium carbonate ($CaCO_3$) or silica (SiO_2). Mesohyl

Fig. 5.27 Sponges in the Caribbean Sea, Cayman Islands: the yellow tube sponge, *Aplysina fistularis*, the purple vase sponge, *Niphates digitalis*, the red encrusting sponge, *Spiratrella coccinea*, and the gray rope sponge, *Callyspongia* sp. Image from the U.S. National Oceanic and Atmospheric Administration

is flanked by layers of living cells. The inner lining consists of choanocytes—"collar cells" with one flagellum each that move water through the pores with whip-like motions. Amebocyte cells transfer food particles along through the sponge. The outer lining of mesohyl contains pinacocytes—cells that can digest food particles; pinacocytes at the bottom of a sponge anchor it in place.

Sponges are members of the phylum Porifera—"pore-bearing" in Latin. Numerous small pores (called ostia) are connected to a network of canals and to larger holes called oscula. Martins et al. (2021) studied North Atlantic deep-sea sponges and found they all have porosities over 68%.

Many sponges are shaped like a tube called an asconoid (Fig. 5.28). Few exceed 1 mm in size; this is because larger sponges need more food and oxygen (as volume, mass and energy requirements scale as the cube of length) whereas the amount of food that can be obtained depends on the surface area covered with choanocytes (which scales as the square of length). In other words, large asconoid sponges would have a surface/volume ratio too high for efficient functioning. Some "syconoid" sponges have modified the asconoid structure by adding internal pleats covered with choanocytes; this increases the exposed surface area and increases the surface/volume ratio.

A strategy adopted by larger "leuconoid" sponges (some exceed 1 m in diameter) is to have an interior filled with porous mesohyl lined with choanocytes; these cells can trap food particles passing through the network of water channels. Leuconoid sponges can assume a variety of shapes including barrels and leaf-like encrusting

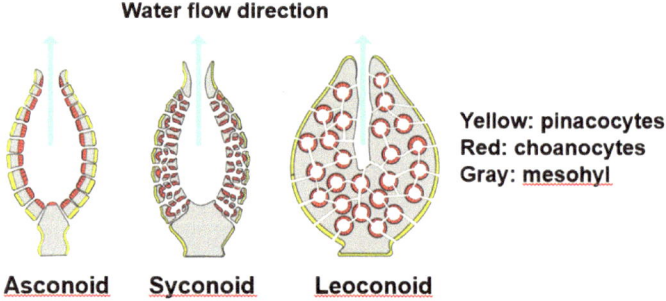

Water flow direction

Yellow: pinacocytes
Red: choanocytes
Gray: mesohyl

Asconoid   Syconoid   Leoconoid

Fig. 5.28 Distinct body structures of sponges showing the direction of water flow, the non-living mesohyl matter and the primary types of living cells. *Image* modified from Philcha

sponges that conform to the underlying rock surface. These sponges all have high surface/volume ratios.

Santavy et al. (2013) examined a variety of sponge forms. For globe-shaped sponges such as *Iricinia strobilina* and *Spheciospongia vesparium*, these researchers determined the median height (h) to be 10 cm and the median diameter (d) to be 15 cm. Because these sponges most closely conform to an oblate ellipsoid, the volume can be estimated from the equation $V = 4/3 \pi a^2 b$ (where **a** and **b** are the minor axes). In the case of the globe-shaped sponges, $b = 1/2h$, and $a = 1/2d$. The calculated volume is ~ 1180 cm^3.

It is difficult to measure the total surface area of a highly porous object, but it can be estimated. A globular sponge could be modeled as a set of spherical shells surrounding spherical pores that are mutually conformable (with no spaces between adjacent shells). If we assume the sponge porosity is 70%, we can model the entire sponge to be a single sphere with a diameter of 12 cm (r = 6 cm) surrounding a hollow sphere with a volume equal to 70% of the volume of the larger sphere. The volume of the large sphere is $4/3 \pi R^3$ or 905 cm^3. The volume of the small hollow sphere is 70% of this value or 634 cm^3; the volume of the spherical shell is 905–634 cm^3 or 271 cm^3. The radius of the small sphere $r = \sqrt[3]{(¾V/\pi)}$, equivalent to 5.33 cm. The surface area of the spongy spherical shell equals the surface area of the large sphere plus the surface area of the small sphere: $4\pi R^2 + 4\pi r^2$ or ~ 810 cm^2. The sponge (modeled as a spherical shell) has a surface/volume ratio of 810/271 = 3 cm^{-1}. If pores were ignored, a spherical sponge of diameter 12 cm would have a surface/volume ratio of 0.5 cm^{-1}. By this crude estimate, a globular sponge has a surface/volume six times greater than a non-porous sphere of the same diameter. These high ratios facilitate the ingestion of food particles and oxygen, allowing larger sponges to thrive.

5.12 Leaves and Trees

5.12.1 Leaves

Robert Frost's (1923) Pulitzer-Prize-winning collection of poems, *New Hampshire*, contains the seemingly quotidian autumnal poem, "Gathering Leaves". The first stanza characterizes the physical properties of dried leaves:

Spades take up leaves
No better than spoons,
And bags full of leaves
Are light as balloons.

In the fifth stanza, Frost describes his shedful of dried leaves as having "next to nothing for weight". Individual leaves must be very light indeed.

Among the trees shedding leaves that Robert Frost had to contend with was the red oak (*Quercus rubra*). These trees are native to southeastern and south-central Canada and the eastern and central United States. Mature red oak trees are commonly 15–25 m tall; some trees exceed 30 m. Trunk diameters are typically 50–100 cm. The upper part of the tree, known as the crown, can be round to egg-shaped. The irregularly furrowed bark is medium to dark gray. The fruits of the red oak are egg-shaped acorns about 5 cm tall; they mature about 18 months after pollination occurs.

Red-oak leaves are arranged alternately on either side of the twig; the oval-shaped leaves have five to eleven lobes and are commonly 13–23 cm long and 6–15 cm wide. The space between the lobes is rounded or U-shaped; lobe tips are sharply pointed with protruding bristles. The upper surface of the leaves is dark green; the lower surface is paler. The petiole (the stalk that joins the leaf to the twig) is slender and commonly 3–5 cm long. The major external parts of a typical leaf are shown in Fig. 5.29.

Average red-oak leaves have a surface area of ~ 50 cm^2 and a specific leaf area (SLA), petiole excluded, of ~ 17 mm^2 mg^{-1}, (Kattge et al., 2020), indicating an individual leaf mass of ~ 300 mg (i.e., about 0.3 g). [About 1500 intact leaves would weigh one pound (0.454 kg).] The leaf thickness (averaged for trees in both understory and open locations) is 0.158 mm (Abrams & Kubiske, 1990). [Note that these values are from different data sets and that leaves from a particular species of tree vary in their properties depending on the amount of available sunlight, moisture, nutrients, mean temperature, etc. These calculations should be viewed only as approximations.] Because red-oak leaves are so thin, the total surface area is approximately double that of one side, i.e., ~ 100 cm^2. The red-oak leaf volume (the surface area of one side times the thickness) is ~ 0.8 cm^3. The leaf density (equivalent to mass divided by volume) is thus ~ 0.38 g cm^{-3}. Because water has an appreciably greater density (1 g cm^{-3}), fallen leaves can float on water (Fig. 5.30).

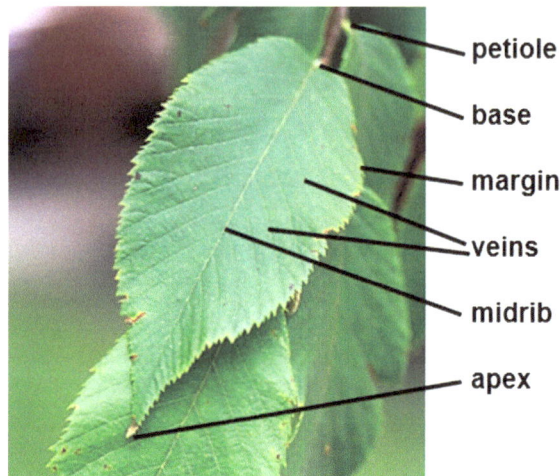

Fig. 5.29 Major external parts of a leaf

Fig. 5.30 Leaves floating on a stream

The surface/volume ratio of the average red-oak leaf is ~ 125 cm^{-1}. In comparison, a sphere with a diameter of 20 cm (roughly the same length as the average cross-section of a red-oak leaf) will have a surface/volume ratio of 0.3 cm^{-1}; this is more than 400 times smaller than that of the typical red-oak leaf.

A cross section of the leaf reveals its internal structure (Fig. 5.31). Both the top and bottom of the leaf are covered with a single transparent layer of cells—the epidermis. The epidermal cells secrete a waxy substance called cutin that forms a water-repellant

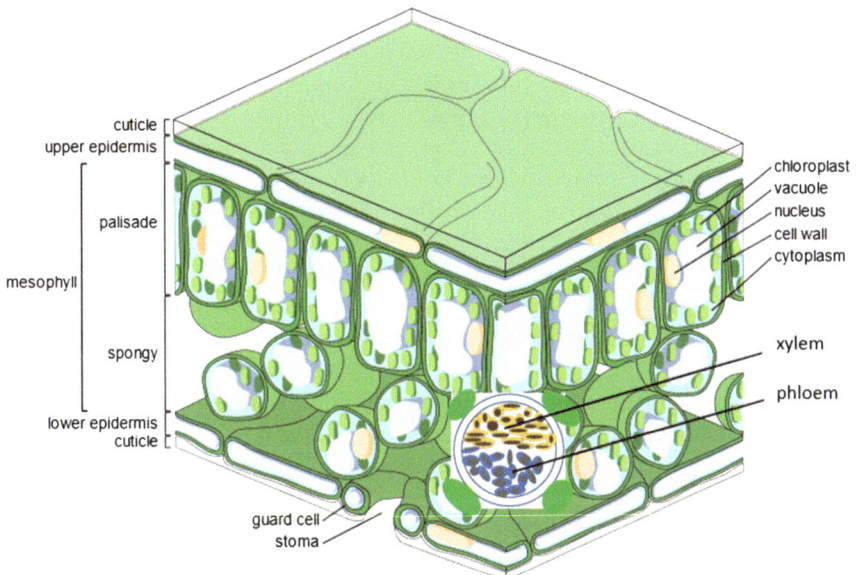

Fig. 5.31 Internal structure of a leaf

layer (the cuticle), a few micrometers thick, along the uppermost surface of the leaf. This layer minimizes leaf dehydration.

Within the epidermis are small pores called stomata (the plural of stoma) that control the rate of gas exchange with the surrounding environment. Stomata are more abundant on the underside of the leaf than near the top. When they are open, stomata permit carbon dioxide to enter the leaf while water vapor and molecular oxygen are expelled. The stomata are flanked by a pair of specialized cells known as guard cells. When the guard cells swell into mirror-image, concave, crescent-shaped masses, the stomata open, permitting gas exchange; when the guard cells shrink, the stomata close and gas exchange is halted.

Between the upper and lower epidermis are two distinct layers of mesophyll cell (Greek for "middle leaf"): a palisade layer (consisting of tightly packed, subparallel, vertically elongated cells, perpendicular to the leaf surface) overlying a spongy layer (of rounded, loosely packed cells separated by air pockets). The spongy layer allows leaves to bend in the wind, reducing drag. Although mesophyll cells all contain chloroplasts—the organelle that conducts photosynthesis (the transformation of carbon dioxide and water into sugar and molecular oxygen)—chloroplasts are more abundant in the palisade cells (those near the upper surface of the leaf and closer to incident sunlight). Chlorophyll occurs within the chloroplasts; it is this chemical that captures energy from sunlight to conduct the photosynthetic reaction. The two most common types of chlorophyll (a and b) have the chemical formulae $C_{55}H_{72}O_5N_4Mg$ and $C_{55}H_{70}O_6N_4Mg$, respectively. Plants derive their green color from chlorophyll because this chemical does not absorb the green wavelengths (~550 nm) of incident white light; these wavelengths are reflected instead.

Within the spongy mesophyll layer is a network of veins containing xylem and phloem; these are the two types of vascular tissue that transport water from the roots of the tree into the leaf (the role of *xylem*) and the photosynthetically manufactured carbohydrates (sugar) from within the leaf to the rest of the tree (the role of *phloem*). Evaporation of water from the leaf (the process known as transpiration) produces a negative pressure, inducing xylem to pull water upwards from the roots via capillary action—a principal mechanism for getting water into the leaf.

Why are these leaves shaped the way they are—with broad flat surfaces, thin flanks, and high surface/volume ratios? Because it is this shape that makes photo-synthesis efficient. The broad flat shape of a leaf facilitates the interception of a lot of sunlight; the thin interiors of the leaves permit the light to penetrate the surface and readily enter the chloroplasts.

Evergreen trees (a category that includes most conifers) also lose their leaves, but at a slow rate. Individual leaves from the bristlecone pine (*Pinus longaeva*), for example, can last 20–30 years. The number of grams of carbon per gram of leaf per year needed for leaf construction is lower for evergreen leaves due to their relative longevity. Evergreen leaves are typically smaller and thicker than deciduous leaves, with a higher leaf-mass per unit of leaf-area. They also have lower nutrient levels per gram and built-in chemical defenses. These characteristics make evergreen leaves tougher to chew, harder to digest, less nutritious, and more repellant to herbivores.

5.12.2   Deciduous Trees

In Sonnet 73, published in 1609, Shakespeare described the autumnal phenomenon of trees shedding their leaves. The first quatrain (presumably spoken by an aging narrator) reads:

> *That time of year thou mayst in me behold*
> *When yellow leaves, or none, or few, do hang*
> *Upon those boughs which shake against the cold,*
> *Bare ruin'd choirs, where late the sweet birds sang.*

This annual leaf shedding is why the season is called "Fall". The term appears to have been coined in England in the Late 16th Century and taken up by North Americans a few decades later. These days, "Fall" is used commonly by residents of Canada and the United States; UK residents tend to call the season "Autumn". The original English name for the season is "harvest".

In wintertime, the Sun is low in the sky and its rays hit the Earth obliquely. The same amount of solar energy is spread over a greater surface area than in the summer, so that the number of photons impinging on each square centimeter of a leaf's surface is diminished. There is also less daylight in each 24-h period; in New Hampshire, for example, there are 6.33 fewer hours of daylight on the winter solstice than on the summer solstice—when the Sun is highest in the sky. These two effects combine to make photosynthesis far less productive in autumn and winter.

The efficiency of photosynthesis is temperature dependent; it is most efficient at temperatures in the range of 20–35 °C (68–95°F). Average daytime high temperatures in New Hampshire in the summer are 21.1–29.4 °C (70–85°F), the peak range of photosynthesis efficiency. Although temperatures drop during the autumn and winter, the sunlight striking a leaf can increase the leaf temperature by 10–20 °C above that of ambient air (Hirons & Thomas, 2018), allowing photosynthesis to continue. When temperatures dip below freezing, however, photosynthesis is inhibited. New Hampshire winters have average low temperatures of −20 and −9.4 °C (−4 to + 15°F). At such low temperatures, liquid water (one of the essential reactants in the photosynthesis process) is essentially unavailable.

Only a percent or two of the water taken up by the tree roots is normally used by the plant for metabolism; roughly 98% of the water is lost through leaves by transpiration. During autumn and winter, leaves need to close their stomata to avoid such drastic loss of water, but this very process cuts off the ability of leaves to absorb carbon dioxide. [Recall that the photosynthesis reaction is: $6CO_2 + 12H_2O$ + light-energy → $C_6H_{12}O_6 + 6O_2 + 6H_2O$.]

Without the ability to conduct photosynthesis, leaves are no longer an asset to the tree. Their high surface/volume ratios would cause extensive loss of water from those stomata that remained open. Wind gusts during winter storms would also cause significant drag on the tree, enhancing the chances of branch damage or complete uprooting. Deciduous trees solve this problem by shedding their leaves (Fig. 5.32). (The word "deciduous" is derived from the Latin *decidere*, meaning to "fall down".)

The daily amount of sunlight in the Northern Hemisphere decreases after the summer solstice (around June 21st). After the autumnal equinox (around September 22nd), each succeeding day sees more darkness than daylight. Trees reduce their production of chlorophyll, allowing the remaining red, orange, and yellow pigments to show. These colors (known as carotenoids) are always present in leaves but are normally overwhelmed by the green color of chlorophyll. The chlorophyll compounds are degraded; most of the nutrients are extracted by the tree. (Carotenoids are degraded more slowly than chlorophyll.) The nitrogen extracted from chlorophyll is transferred to other tissues of the tree where it is used to manufacture amino acids, nucleic acids, and proteins.

Emily Brontë summarized the relationship between falling leaves and the diminishing hours of daylight that characterize the autumn season:

Fall, leaves, fall; die, flowers, away;
Lengthen night and shorten day;
Every leaf speaks bliss to me
Fluttering from the autumn tree.

Each autumn, a gray-green-colored protective layer of cells, called the abscission layer (Fig. 5.33), grows between the leaf stem (the petiole) and the tree branch. This layer serves as a barrier, preventing the transportation of additional water and nutrients to the leaf. The abscission zone is actually composed of two layers in

Fig. 5.32 An American elm tree ("The Grayson Elm") in Amherst, Massachusetts. The 24-m-tall tree is full of leaves in mid-summer (August 2017) and denuded of leaves in mid-winter (January 2013). *Photographs* by Msact at English Wikipedia

deciduous trees: (1) a top layer containing cells with weak walls and (2) a bottom layer that absorbs a lot of water and begins to swell. The expansion of the bottom layer breaks the weak cell walls in the top layer, causing the leaf to fall off.

Fig. 5.33 Abscission layer of a leaf in autumn. *Image* courtesy of the Public Library of Science

5.13   Human Anatomy

5.13.1   The Brain

Although the average adult human brain weighs 1200–1500 g (only ~ 2% of total body weight), it uses about 20% of the body's energy—more than any other organ. Such energy expenditure reflects the myriad tasks the brain is engaged in, most beyond conscious control. The brain is the locus of thought, emotion, and memory, the site of abstract reasoning and problem solving, the processor of information from our sense organs, the manager of the body's breathing, the regulator of its temperature, the controller of muscle movement, the producer of speech, and the instigator of sensations of hunger, thirst, sexual desire, and the urge to head for the toilet.

There are three main parts of the brain (Fig. 5.34)—the cerebrum (the largest portion, constituting 85% of the brain's weight), the cerebellum (a fist-sized mass at the back of the head), and the brainstem (a stalk connecting the brain to the spinal cord). The brain, together with the spinal cord, make up the central nervous system.

Cerebrum. This portion of the brain provides instructions for body movement and temperature regulation; it is the site that receives and processes sensory information; it facilitates speech, reasoning, and learning, and governs our emotions. The outer 2–4-mm-thick region of the cerebrum consists of gray matter—the cerebral cortex; it is composed of rounded, bulbous, neuron somas (cell bodies) that contain the cell nucleus. This is the site where incoming information is processed and interpreted. The interior of the cerebrum consists of white matter—long stems called axons that connect different neurons. Axons are wrapped in sheaths of myelin (a protective coating consisting of protein and fat). They transmit information from neuron somas to other regions of the central nervous system.

The cerebrum is divided into two hemispheres—the right hemisphere controls the left side of the body, and the left hemisphere controls the right side. Between

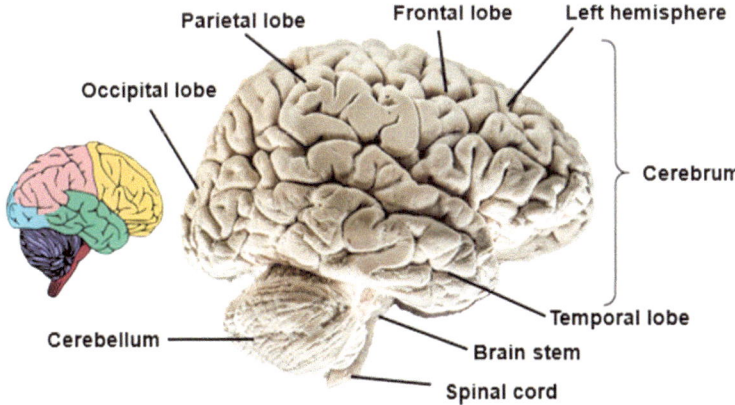

Fig. 5.34 Major parts of the human brain

the hemispheres is an indentation known as the medial longitudinal fissure. The two hemispheres exchange information through the corpus callosum. Each hemisphere consists of four lobes—the frontal lobe (which coordinates executive functions including self-restraint, reasoning, abstract thought, personality characteristics, smell, and (via Broca's area) speech), the temporal lobe (involved with memory, emotion, auditory stimuli, speech, smell, and musical rhythm), the parietal lobe (responsible for processing sensory information from the body and (via Wernicke's area) enabling the understanding of spoken language), and the occipital lobe (associated with vision).

Cerebellum. The cerebellum is the region of the brain in charge of the body's posture, equilibrium, and balance; it is the site where voluntary muscle movements are coordinated. It was recently found also to be involved in emotion, behavior, and cognition. Like the cerebrum. the cerebellum is divided into left and right hemispheres.

Brainstem. The brainstem, which joins the cerebrum to the spinal cord, is made up of the midbrain (which enables hearing and body movement and the body's responses to environmental conditions), the pons (which facilitates chewing, hearing, visual focus, facial expression, and tear production), and the medulla (the regulator of breathing, heart rate and blood flow, as well as the body's levels of oxygen and carbon dioxide). The medulla connects directly to the spinal cord.

A glance at Fig. 5.34 shows the cerebrum is characterized by numerous folds. The folds themselves consist of ridges (called gyri) and troughs (called sulci). Although the cerebrum is much larger than the cerebellum, the cerebellum is more tightly folded, accounting for 78% of the total surface area of the brain (Sereno et al., 2020).

The average adult woman has a brain volume of 1130 cm^3 (e.g., Cosgrove et al., 2007) and a brain total surface area of 2038 cm^2 (Herron et al., 2015), leading to a surface/volume ratio of 1.8 cm^{-1}. If the brain were smooth, the surface/volume ratio would be significantly lower; a sphere with a volume of 1130 cm^3 would have a surface/volume ratio of 0.46—about four times lower than the average female-human brain.

The numerous folds in the human brain dramatically increase its surface/volume ratio, thereby permitting a great deal of cognitive functioning to occur within a relatively small volume. If it were not for these folds, we would either be less cognitively able or forced to walk around with large, unwieldy heads (Fig. 5.35).

5.13.2 The Respiratory Tract

The respiratory tract (Fig. 5.36) is divided into two portions: (1) the upper tract, consisting of the nose, nasal cavities, sinuses, pharynx, and upper part of the larynx, and (2) the lower tract, consisting of the lower part of the larynx, the trachea, and the lungs (and their constituent parts). Like most vertebrates, humans have two lungs located in the thoracic cavity in the chest, near the backbone, beneath the rib cage, flanking the heart. Below the lungs is the diaphragm, the principal muscle involved in

respiration; contraction of the diaphragm during inhalation expands the chest cavity. The left lung is the smaller of the two; it has a concave indentation, called the cardiac notch, against which the heart rests. In addition to its principal function of respiration, the lungs also provide the airflow necessary for human speech.

Each lung is divided into lobes (three on the right, two on the left) by a thin membrane called the visceral pleura. Each lobe is divided by fissures (infoldings of this membrane) into discrete segments. Lung segments are divided into lobules. Secondary lobules (typically 1–2.5 cm in diameter) are the smallest portions of the lung, separated from one another by connective tissues containing elastic fibers. Each secondary lobule contains 30 or so primary lobules.

The base of the trachea is divided into two main bronchi—the right main bronchus (supplying air to the right lung) and the left main bronchus (supplying air to the left lung). These primary bronchi are the widest airways in the respiratory tract; at the sites where these bronchi enter the lung, they branch into secondary bronchi, which bring air to each lobe. Secondary bronchi branch into tertiary bronchi (a.k.a. segmental bronchi) that conduct air to the segments. Tertiary bronchi are divided and subdivided into fourth-, fifth- and sixth-order bronchi, all of which supply air to segments of the lungs. The bronchi are kept open by incomplete rings and plates of cartilage.

Sixth-order bronchi subdivide into bronchioles, typically of millimeter size. These narrow passageways do not contain cartilage but are instead lined with muscle tissue.

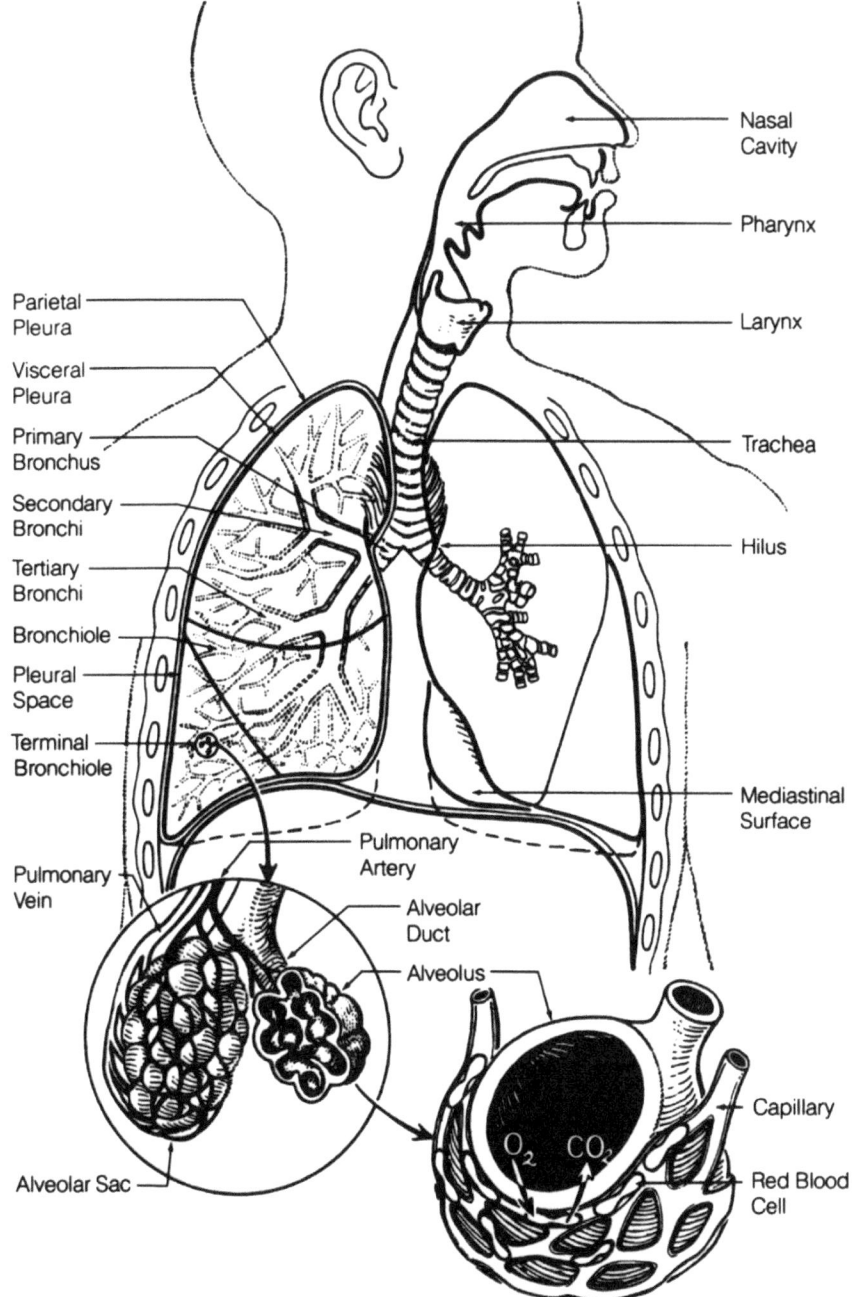

Fig. 5.36 The respiratory system

The diameters of the bronchioles change, allowing air flow to increase or decrease. The smallest bronchioles are called terminal bronchioles (≤ 0.5 mm in diameter); there are a few dozen terminal bronchioles per bronchiole. The terminal bronchioles are the last portions of the conducting zone. (This zone is the upper portion of the respiratory tract that moves gases into and out of the lungs. It is not involved in gas exchange.) The set of branching bronchi and bronchioles resembles a fractal tree; the pattern of air passages appears essentially the same at different levels of magnification.

The two lungs together contain roughly 2400 km of airway passages.

Terminal bronchioles branch into respiratory bronchioles; these are lined mainly with a thin tissue called epithelium. The respiratory bronchioles constitute the first portions of the respiratory zone. This zone is composed of three components (Fig. 5.37): (1) the respiratory bronchioles, (2) alveolar ducts (side branches of the respiratory bronchioles), and (3) alveoli (air sacs in which oxygen is exchanged for carbon dioxide). It is this zone that is responsible for exchanging inhaled molecular oxygen (O_2) for carbon dioxide (CO_2). The oxygen diffuses into capillaries in the lungs and enters the blood stream.

Respiratory bronchioles (≤ 0.5 mm across) conduct air to the alveoli, small cavities in the lungs where gas exchange takes place.

Each alveolus is 200–500 µm in size; approximately 70% of each alveolus is enmeshed in a network of capillaries. There are about 300–500 million alveoli in the lungs, creating a huge surface area: about 75 square meters. Numerous elastic fibers occur in the alveoli, permitting them to expand during inhalation. During expansion, the surface area is increased, enhancing diffusion. Relatively few alveoli are present

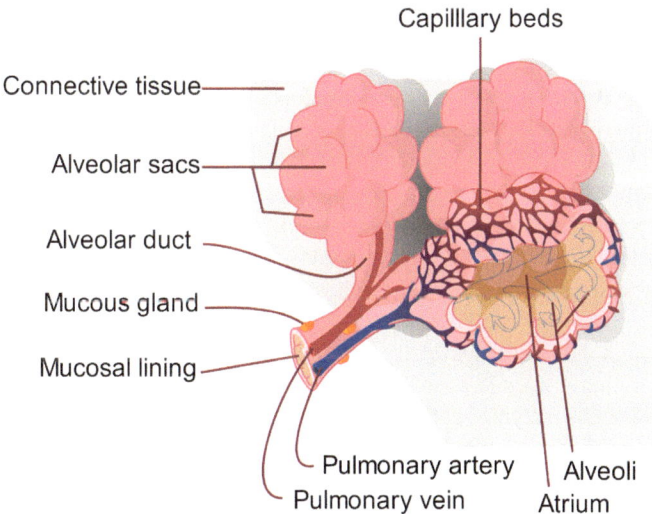

Fig. 5.37 Pulmonary alveoli are hollow cavities in which the respiratory bronchioles exchange molecular oxygen for carbon dioxide with the bloodstream

in the respiratory bronchioles close to the terminal bronchioles; their concentrations greatly increase farther down the line.

Alveolar ducts branch off the sides of the respiratory bronchioles. Five or six individual alveolar sacs sit atop these ducts like bunches of grapes; the sacs are lined with large numbers of alveoli. The endpoint of this pathway (consisting of the respiratory bronchioles, alveolar ducts, alveolar sacs, and alveoli) is called an acinus.

The pair of lungs in adults has a total weight of about 1000–1300 g and a volume of roughly 5000–6000 ml, where 1 ml is equivalent to 1 cm³. Because alveoli constitute about 90% of total lung volume, it is reasonable to assume their total volume is also roughly 6000 ml and that the total surface area of the lungs is equivalent to that of the alveoli: ~ 75 m². Thus, the surface/volume ratio of the lungs is about 125 cm^{-1}. In contrast, the surface/volume ratio of a sphere of equivalent lung volume (6000 cm³) would be 0.27 cm^{-1}. This value is roughly 460 times lower than the surface/volume ratio of the lungs.

To summarize, it is the very high surface/volume ratio of the lungs, caused by the presence of hundreds of millions of tiny alveoli, that permit efficient gas exchange and allow proper lung function.

5.13.3 The Gastrointestinal Tract

The gastrointestinal tract (a.k.a. the alimentary canal) is the course through which food is ingested and digested and solid waste is expelled. It consists of the mouth, pharynx, esophagus, stomach, small intestine, large intestine, and anus. About 10% of the body's energy is expended by digestion. The entire process is highly efficient; generally, less than 10% of ingested food is expelled as waste.

Chewing is the initial step in the digestive process. It involves grinding and crushing the food by the teeth, making food softer and warmer and markedly increasing the food's surface/volume ratio. This facilitates the chemical breakdown of the food by enzymes in the saliva; when food particles are small, enzymes have a lot of available surface to interact with and only minimal depths to penetrate.

The pharynx, located in the throat behind the mouth, allows food to enter the esophagus and air to enter the larynx. (A wing-shaped flap called the epiglottis prevents food from getting into the trachea.)

After the chewed, partially broken-down food (now called a bolus) is swallowed, it travels through the esophagus where it is guided toward the stomach by peristalsis— radially symmetric muscle contractions and relaxations. After the bolus enters the stomach, digestive enzymes and gastric acid are secreted, facilitating the breakdown of long chains of amino acids in the ingested proteins. The degree of acidity (the pH) of the gastric acid is regulated by the production of bicarbonate (HCO$^-_3$) by some stomach cells. A muscular valve, called the pylorus, is the gateway for this mixture of partially digested food and gastric juices (now called chyme) to enter the small intestine (Fig. 5.38).

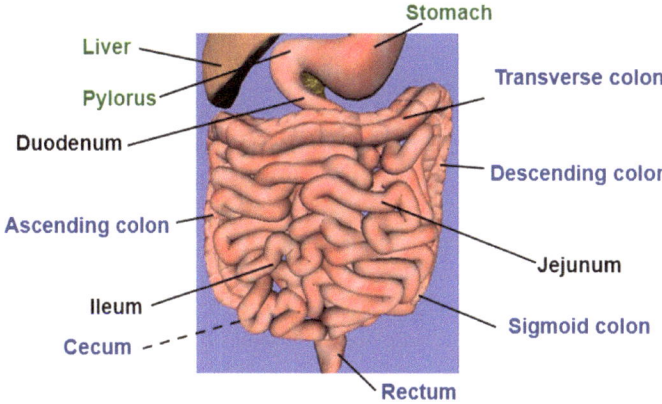

Fig. 5.38 The small intestine, composed of the duodenum, jejunum, and ileum, is flanked by the large intestine. The cecum (the first section of the large intestine) is indicated by a dashed line because it is hidden behind a portion of the small intestine in this diagram. Also shown are some adjacent organs

The small intestine is composed of three parts—the duodenum (the smallest component, adjacent to the pylorus), the jejunum (the central component), and the ileum (the longest component, the one that delivers the remaining food to the large intestine). There is a thin membrane called the mesentery that extends from the posterior of the abdominal cavity to envelop the pancreas, jejunum, and ileum, anchoring these organs in place.

Duodenum. This is the shortest portion of the small intestine; it is generally 20–25 cm in length. It receives chyme from the stomach along with digestive enzymes from the pancreas (which break down proteins) and bile from the liver (which breaks down fats). Glands in the duodenum manufacture a bicarbonate-rich mucusy secretion that combines with bicarbonate from the pancreas to buffer the acidity of the chyme.

Jejunum. This middle portion of the small intestine is typically about 2.5-m long. It connects the duodenum to the ileum and is characterized by numerous circular folds. Within the lining of the jejunum are intestinal villi—millimeter-size cylindrical stalks that greatly increase the surface area of the intestinal interior. There are about 3000 villi per square centimeter. Each villus contains numerous protruding microvilli; by one estimate, there are roughly 20 billion microvilli per square centimeter along the intestinal lining. These villi and microvilli together have extremely high surface/volume ratios; this greatly facilitates absorption into the bloodstream of the nutrients (sugars, amino acids and fatty acids) liberated from digested food.

Ileum. This last portion of the small intestine averages about 3 m in length. The interior of the ileum is also lined with intestinal villi. The ileum absorbs vitamin B_{12} (cobalamin), bile acids produced in the liver, and residual nutrients.

The average small intestine in an adult is roughly 2.8 cm in diameter and about 5.5 m long. Numerous folds allow this organ to fit into a rather compact space within the abdominal cavity. We can use these dimensions to calculate the volume and surface area of the small intestine (excluding the presence of villi and microvilli) by assuming the small intestine has been straightened out and has the shape of a cylinder:

$$\text{surface area } (SA = 2\pi rh + 2\pi r^2) \text{ and volume } (V = \pi r^2 h)$$

We obtain a surface area of 4850 cm^2, a volume of 3387 cm^3, and a surface/volume ratio of 1.43 cm^{-1} for the average small intestine. The surface/volume ratios of a sphere and cube of equivalent volume would be 0.32 cm^{-1} and 0.40 cm^{-1}, respectively; these values are 4.5 times and 3.6 times smaller than that of the small intestine. The high surface/volume ratio of the small intestine is an essential feature that allows efficient chemical breakdown of food and the absorption of the nutrients necessary to power the body.

After leaving the ileum, the remaining mixture of food and digestive juices (mainly plant fiber, dead intestinal cells, water, salt, and bile pigments) enters the large intestine. This organ is called the *large* intestine because it is appreciably wider than the *small* intestine (~4.8 cm vs. ~ 2.8 cm in diameter), even though it is much shorter (~1.5 m vs. 5.5 m). The first portion of the large intestine is the cecum, a 5-cm-long tube separated from the ileum by a valve. The cecum absorbs fluids and salts, mixes undigested food with mucus, and passes along the mixture to the colon.

The colon is the major portion of the large intestine; it consists of four segments— the ascending colon, transverse colon, descending colon, and sigmoid colon. Water and nutrients are removed throughout the colon. The residual food material is essentially waste; it moves upward through the ascending colon by peristalsis, moves across the transverse colon, and then down the descending colon. This is the part of the colon where feces are stored before being emptied into the rectum. The sigmoid colon derives its name from its shape—it looks like the Greek letter *sigma*, σ. Contraction of the muscles in the sigmoid colon pushes feces into the rectum—the final, 12-cm-long, portion of the large intestine. From here, stool is pushed into the anal canal; this 4-cm-long channel, terminating at the anus, regulates the elimination of waste via the contraction of sphincter muscles. At times, this has been a public activity (Fig. 5.39). So it goes.

Fig. 5.39   Public latrines in Ostia Antica, the harbor city of ancient Rome. At the time, the city was at the mouth of the River Tiber

5.14   Human Anatomical and Behavioral Responses to Extreme Temperatures

When the weather is hot, humans (and horses) can sweat profusely to cool down. Eccrine sweat glands[1] in the skin secrete brackish water that can be lost by evaporative cooling, lowering the body temperature. These glands occur all over the human body, with the densest concentrations occurring on the palms and soles. Concentrations are somewhat lower on the head and appreciably lower elsewhere on the body. To prevent overheating, people can also seek shade, wear protective clothing (Fig. 5.40), or restrict vigorous activity to cooler hours of the day.

When the weather is cold, human bodies can shiver (or shudder). This is an involuntary reflex caused by the shaking of skeletal muscles, producing heat. Goose bumps (a.k.a. goose flesh or goose pimples) can form in response to cold. Arrector pill muscles in humans and non-human apes are connected to hair follicles beneath the skin; when these muscles contract (a reflex known as piloerection) the body hair (or fur) is raised. This can make the body look larger (due to an increase in surface area) and, thus, more formidable to potential predators or enemies. Raising the hair also increases the volume of air near the skin, forming an insulating layer. Water droplets clinging to the strands of hair are raised up away from the skin; when these

---

[1] There is a second type of sweat gland—the apocrine gland—that occurs mainly on the armpits and perineal areas (between the anus and external genitalia). It secretes an oily substance whose (body) odor is due to bacterial activity. Modified apocrine sweat glands include the milk-producing mammary glands, ear-wax-producing ceruminous glands, and sebum-producing ciliary glands in the eyelids.

Fig. 5.40 Protective clothing worn in the Sahara. Detail of an image by Sergey Pesterev of a caravan in the dessert, Morocco

droplets evaporate, heat is lost from the insulating air layer, rather than from the skin. The body therefore loses appreciably less heat than it would otherwise.

People can also modify their behavior by diminishing their exposure to the cold. They can seek shelter and put on layers of insulating clothing. When at rest, cold individuals tend to curl up in a fetal-like position with curved back, lowered head, and limbs bent and pressed tightly against the torso (Fig. 5.41). Because the volume of the body remains the same, this position diminishes the surface area of the exposed portions of the body, thereby decreasing the body's effective surface/volume ratio. Insulating air is trapped against the body; the surface area of exposed skin is decreased. This results in diminished heat loss.

People huddling together on a cold night seek the same result (Fig. 5.42). Whether they huddle or not, the total human volume of the huddlers is unchanged. Huddling reduces the total skin surface area of the group that is exposed to cold air; this decreases the effective surface/volume ratio of the huddled group. The same wintertime strategy has been adopted by Emperor penguins in Antarctica and garter snakes lying together in some warm cozy spot (a hibernaculum) in the midwestern United States.

5.15   Frostbite

Frostbite can occur when water in the skin freezes into ice crystals due to exposure to cold conditions (cold air or cold water) (Fig. 5.43). It has plagued humanity for thousands of years; it was discussed by the ancient Greeks around 400 B.C.E. (Zafren, 2013). There are three stages in the development of frostbite: (1) Frostnip—the skin may become red or pale and feel cold, tingly, or numb. (2) Superficial Frostbite—the skin may feel warm; it may begin to sting or swell. After a day or two, the skin might peel or blister. (3) Severe Frostbite—the subcutaneous layers of the skin freeze; cell

Fig. 5.41 Cold fellow decreasing the surface/volume ratio of the part of his body exposed to frigid ambient air. *Drawing* by Jeremiah Rubin

Fig. 5.42 U.S. Naval Personnel, wet and cold and covered in mud, after completing a training exercise. They are huddling together, unconsciously reducing their collective surface/volume ratio, to diminish heat loss. *Photograph* taken by Chief Photographer's Mate Chris Desmond and released by the U.S. Navy

membranes and small blood vessels are damaged. After a day or two, large blisters form and the skin turns black. There is an increased risk of infection and gangrene.

The parts of the body that are most likely to become frostbitten are the fingers, toes, and ears. These extremities are long and thin and protrude from the body. They have high surface/volume ratios and are poised to lose heat much faster than the arms, legs, and trunk.

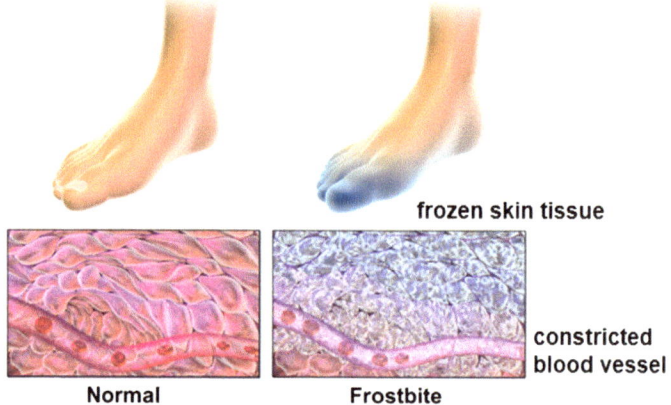

Fig. 5.43 Comparison of normal and frostbitten tissue. *Image* by BruceBlaus

5.16   Bacteria

Bacteria are the most abundant organisms on Earth. They are present in oceans, lakes, and ponds; they occur in soil; they inhabit deserts, rainforests and acidic hot springs; they can be found deep within the Earth's crust and lurking around hydrothermal vents on the ocean floor; they occur in the cores of nuclear reactors and on the surface of unsterilized medical equipment. They form complex symbiotic relationships with other organisms. Many are parasites. Altogether, there are about 10^{30} individual bacteria sharing our planet (Flemming and Wuertz, 2019).

They are also abundant in the human body—a recent study estimates there are about 3.8×10^{13} bacteria (that is 38 trillion bacteria) in the human colon (Sender et al., 2016). There are roughly as many bacterial cells in the human body as human cells. But because bacteria are so small, their collective weight is only about 200 g (i.e., about 0.3% of the total adult human body weight).

Robert Hooke was the first to observe a microorganism, describing the microfungus *Mucor* in his 1665 book *Micrographia*. He coined the term "cell", noting the resemblance of the fungus to a honeycomb. Bacteria were discovered a few years later by Antoni van Leeuwenhoek using a single-lens microscope of his own design. He called these microbes "small animals", translated into English as "animalcules" (Fig. 5.44). Leeuwenhoek wrote numerous letters to the Royal Society in London, describing his discoveries in detail.

Most bacteria are single-celled organisms; all are prokaryotic—meaning their cells lack a nucleus. Bacteria (Fig. 5.45) are encased in a cell membrane composed mainly of phospholipids—mushroom-shaped molecules with a head at one end containing a phosphate group $[(PO_4)^{3-}]$ and a stalk at the other end containing fatty acids joined together with an alcohol (e.g., glycerol; $HOCH_2CHOHCH_2OH$). The cell walls are formed by peptidoglycan (PG)—a macromolecule made of sugars and amino acids surrounding the cytoplasmic membrane.

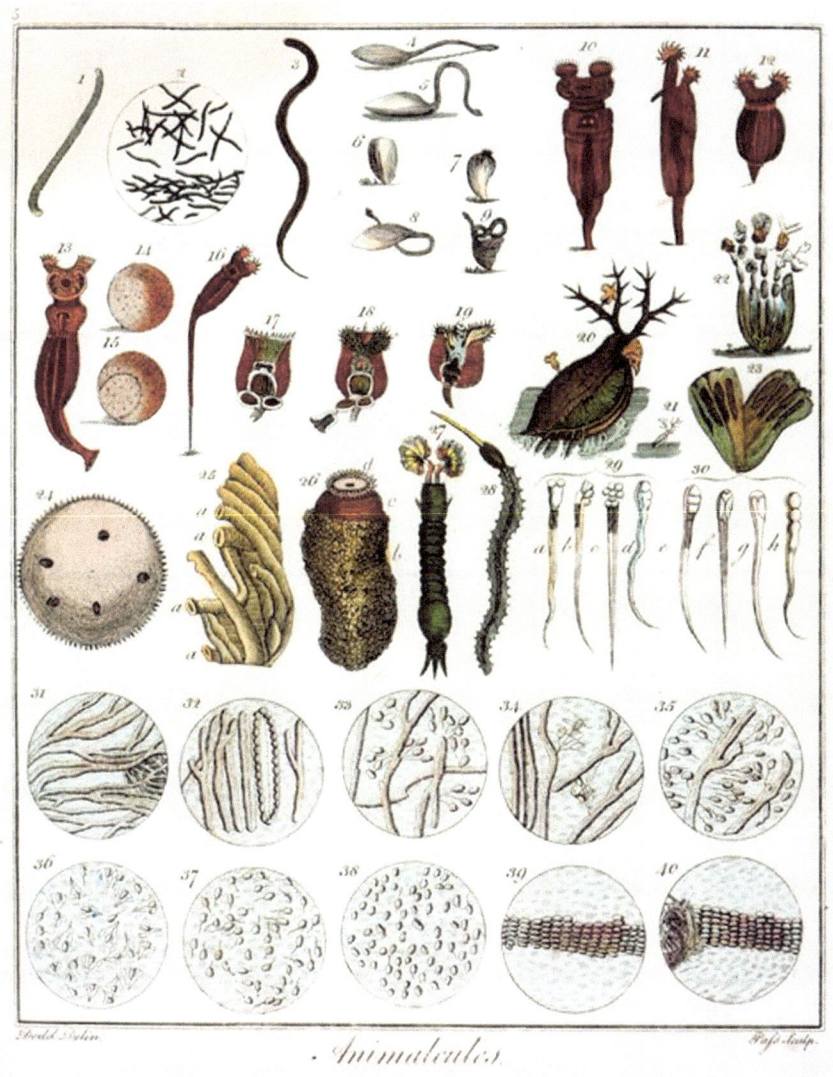

Fig. 5.44 An unknown artist's 1795 rendering of van Leeuwenhoek's animalcules

Bacterial cells are typically 0.5–5.0 μm long, although one giant species, *Epulopiscium fishelsoni*, can exceed 700 μm in size (0.7 mm), easily visible to the naked eye. Recently, and surprisingly, a huge bacterium, *Thiomargarita magnifica*, was described; it consists of a filamentary single cell more than a centimeter long (larger than a female black widow spider).

Most bacteria come in one of three basic shapes (Fig. 5.46)—spheres or ovoids (known as cocci; e.g., *Streptococcus pyogenes*, which causes strep throat), rods

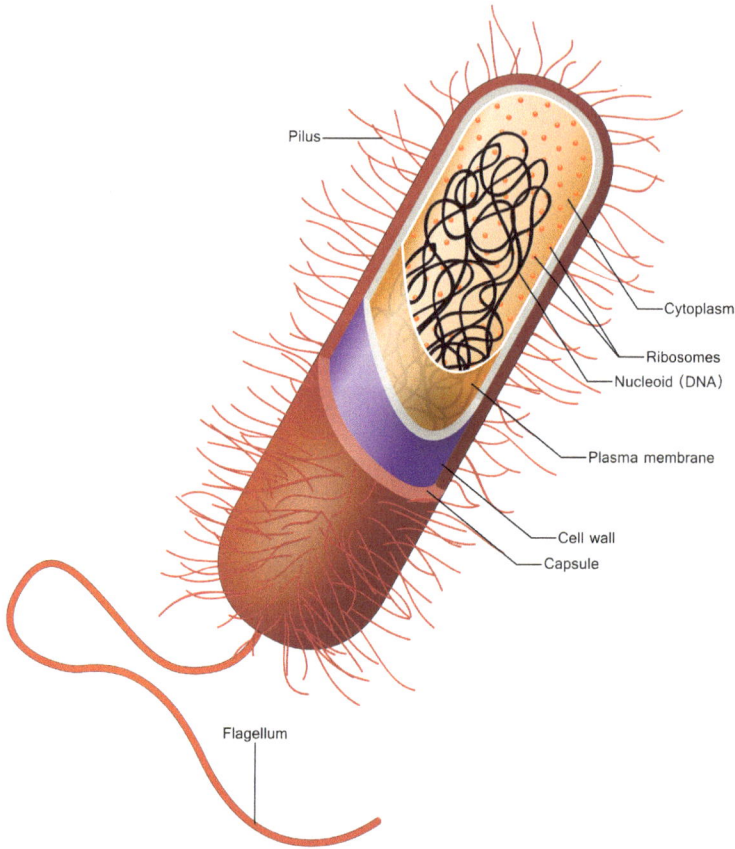

Fig. 5.45 Parts of a rod-like bacterial cell

(bacilli; e.g., *Escherichia coli*, some strains of which can cause food poisoning, septic shock, or meningitis), commas (vibrios; e.g., *Vibrio alginolyticus*, which can cause ear infections), spirals (spirilla; e.g., *Treponema pallidum*, which causes syphilis), or corkscrews (spirochetes; e.g., *Borrelia burgdorferi*, which causes Lyme disease). Some species depart from these basic shapes, e.g., forming filaments (e.g., *Sphaerotilus natans*), boxes (e.g., *Haloarcula marismortui*), or stars (e.g., *Stella vacuolata*). Some bacteria can modify their shape as they grow, particularly in response to environmental stressors.

Colonies of bacteria display a myriad of arrangements. Cocci can occur as unattached single spheroidal cells, as two conjoined cells, as tetrads, cubes, chains, or clusters. Bacilli appear most commonly as individual rods; some arrange themselves in chains like sausages on a string. Following cell division, some bacilli species form palisades, with individual rods stacked side by side like a picket fence.

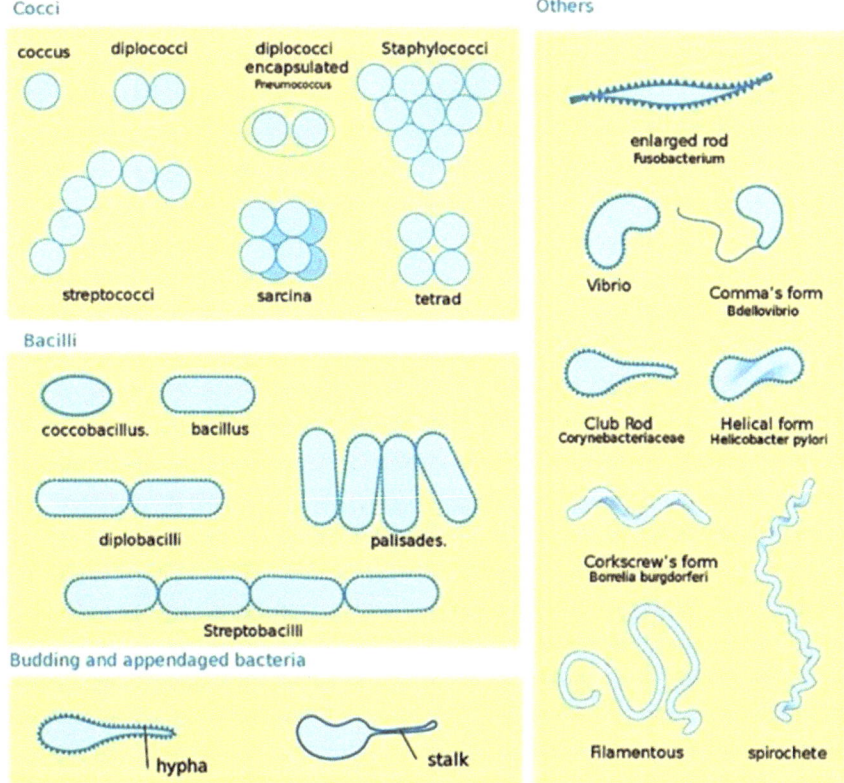

Fig. 5.46 The large variety of bacterial shapes. *Image* modified from original by Mariana Ruiz LadyofHats

Adhering bacteria can aggregate and stick to organic and inorganic surfaces to produce slimy biofilms. The most familiar biofilm (composed mainly of bacteria, but also including fungi) is dental plaque, present on and in between teeth and on and below the gumline.

Some bacteria contain long, thin, hair-like appendages (typically 0.02-μm thick and 20-μm long) called flagella (Fig. 5.47) that use metabolic energy to propel the bacteria forward. (Mammalian sperm cells are similarly flagellated.) Also present in some species are short, hair-like appendages called pili or fimbriae located at the cell surface. These structures enable pathogenic bacteria to adhere to the surface of tissues of infected organisms during colonization.

Harris et al. (2014) studied *Caulobacter crescentus*, a crescent-shaped bacterium that survives on low levels of nutrients in fresh water. They noted that a mutation in MreB (a protein that influences cell growth) causes individual bacteria to develop wide variations in size and shape. This protein became concentrated in certain regions of the cell, enabling rod-like growth; however, this growth was compensated for by

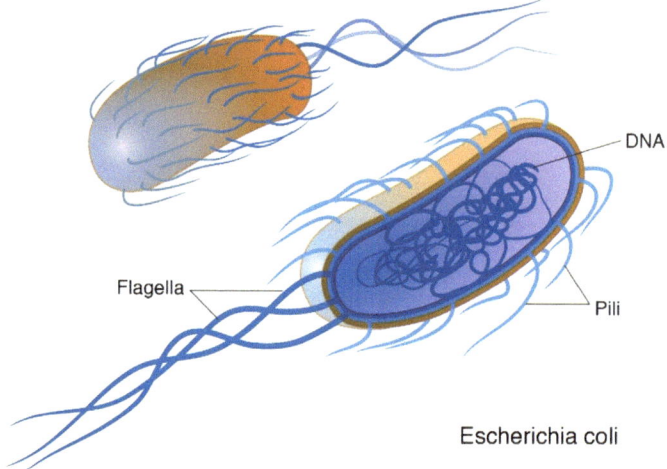

DNA

Flagella

Pili

Escherichia coli

Fig. 5.47 *E. coli* with flagella. *Image* courtesy of the NIH National Human Genome Research Institute

widening in other regions of the cell, so that the cell, as a whole, maintained a relatively constant surface/volume ratio.

In contrast, under growing conditions that promoted increases in the surface/volume ratio, bacterial cell widening was curtailed to maintain this ratio. Follow-up studies by Harris and Theriot (2016, 2018) found that many unrelated bacterial species behave in a similar manner; after being exposed to environmental stressors, they naturally reverted to the surface-area/volume ratio of their wild, unstressed counterparts. They stated that "if the current SA/V is not equal to the ratio of the surface and volume growth rates, cells alter their dimensions until the proper ratio is achieved." It is the biosynthesis of cell walls, mediated by the availability of peptidoglycan (PG), that controls cell morphology. Large, fast-growing cells produce less PG relative to their volume; slow-growing cells show increases in PG production.

Why should many bacteria strive to maintain constant surface/volume ratios? Because the characteristic surface/volume ratio of each species approaches the optimal configuration for absorbing nutrients from the environment.

5.17   Oxygen Diffusion Through Red Blood Cells

Red blood cells (a.k.a. erythrocytes) in vertebrates are responsible for delivering molecular oxygen (O_2) from the respiratory system to the tissues of the body. They contain hemoglobin—the iron-rich protein that can bind and transport oxygen. Human red blood cells are 7.5–8.7 μm in diameter and 1.7–2.2 μm thick (Fig. 5.48).

There are roughly 270 million molecules of hemoglobin in every human red blood cell. The cell is enclosed by a membrane consisting of proteins and lipids; this structure provides stability to the cell during its travels through the circulatory system.

Jones (1979) calculated the surface/volume ratio of red blood cells in seven verte-brates (goat, sheep, horse, rabbit, dog, man, bullfrog) from literature data on cell size after modeling individual cells as cylindrical disks. He also used data from Holland and Forster (1966) on the velocity constants for the initial uptake of oxygen by red blood cells. Jones showed that the smaller red blood cells, i.e., those with higher surface/volume ratios, absorb O_2 more efficiently (Fig. 5.49). As red cells increased in size, there was proportionately less cell surface area. He concluded that "the rate of hemoglobin saturation is limited primarily by the surface area through which oxygen can diffuse in relation to the cell volume…" It is clear that diffusion is faster through the interiors of smaller objects—specifically, those with high surface/volume ratios.

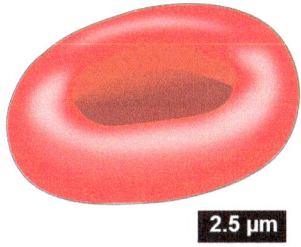

2.5 μm

Fig. 5.48 Illustration of human red blood cell. *Modified image* from Wikimedia Commons

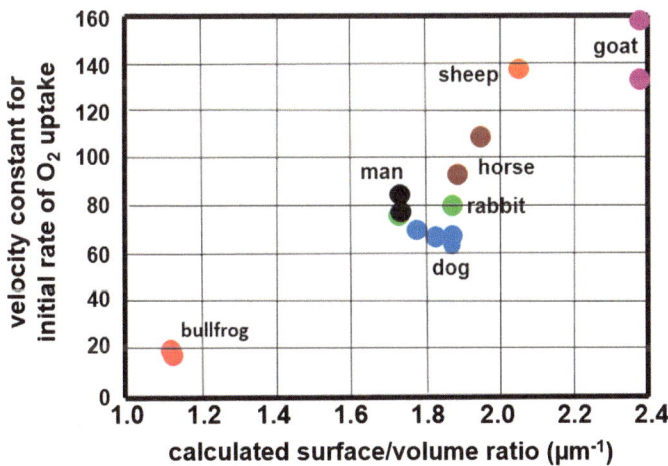

Fig. 5.49 Diagram based on data from Jones (1979) showing the strong correlation in red blood cells in seven vertebrate species between the velocity constant for the initial rate of uptake of molecular oxygen at 37 °C and the calculated surface/volume ratio. Red = bullfrog; blue = dog; green = rabbit; black = man; brown = horse; orange = sheep; purple = goat

References

Abrams, M. D., & Kubiske, M. E. (1990). Leaf structural characteristics of 31 hardwood and conifer tree species in Central Wisconsin: Influence of light regime and shade-tolerance rank. *Forest Ecology and Management, 31*, 245–253.

Aguirre, T. G., Ingrole, A., Fuller, L., Seek, T. W., Fiorillo, A. R., Sertich, J. J. W., & Donahue, S. W. (2020). Differing trabecular bone architecture in dinosaurs and mammals contribute to stiffness and limits on bone strain. *PLoS ONE, 15*, e0237042. https://doi.org/10.1371/journal.pone.023 7042

Alexander, D. E. (2002). *Nature's Flyers: Birds, insects, and the biomechanics of flight* . Johns Hopkins University Press, Baltimore, 358 pp.

Allen, J. A. (1877). The influence of physical conditions in the genesis of species. *Radical Review, 1*, 108–140.

Baum, S. (1997). Hesse's rule. *Glossary of oceanography and the related geosciences with references.* Texas Center for Climate Studies, Texas A&M University

Benedict, F. G. (1934). Die Oberflächenbestimmung Verschiedener Tiergattungen. *Ergebnisse Der Physiologie Und Experimentellen Pharmakologie, 36*, 300–346.

Bergmann C. (1848). *Über die Verhältnisse der Wärmeökonomie der Thiere zu ihrer Grösse.* Vandenhoeck und Ruprecht, Göttingen.

Bonner, J. T. (2006). *Why size matters.* Princeton University Press, 161 pp.

Carroll L. (1865). *Alice's adventures in wonderland.* Illustrations by John Tenniel. Macmillan, London

Cosgrove, K. P., Mazure, C. M., & Stanley, J. K. (2007). Evolving knowledge of sex differences in brain structure, function, and chemistry. *Biological Psychiatry, 62*, 847–855.

Flemming H.-C., & Wuertz S. (2019) Bacteria and archaea on Earth and their abundance in biofilms. *Nature Reviews Microbiology, 17*, 247–260.

Frost, R. (1923). *New Hampshire.* Henry Holt.

Geist, V. (1987). Bergmann's rule is invalid. *Canadian Journal of Zoology, 65*, 1035–1038. https://doi.org/10.1139/z87-164

Haldane J. B. S. (1926, March) On being the right size. *Harper's Magazine.*

Harris, L. K., & Theriot, J. A. (2016). Relative rates of surface and volume synthesis set bacterial cell size. *Cell, 165*, 1479–1492.

Harris, L. K., & Theriot, J. A. (2018). Surface area to volume ratio: A natural variable for bacterial morphogenesis. *Trends in Microbiology, 26*, 815–832.

Harris, L. K., Dye, N. A., & Theriot, J. A. (2014). A *Caulobacter* MreB mutant with irregular cell shape exhibits compensatory widening to maintain a preferred surface area to volume ratio. *Molecular Microbiology, 94*, 988–1005.

Heinrich, B. (1974). Thermoregulation in endothermic insects. *Science, 185*, 747–756.

Hemmingsen, A. M. (1960). Energy metabolism as related to body size and respiratory surfaces, and its evolution. *Report of Steno Memorial Hospital (Copenhagen), 9*, 1–110.

Herron, T. J., Kang, X., & Woods, D. L. (2015). Sex differences in cortical and subcortical brain anatomy. *F1000Research, 4*, 88. https://doi.org/10.12688/f1000research.6210.1

Hirons, A. D., & Thomas, P. A. (2018). *Applied tree biology.* Wiley, 411 pp.

Holland, R. A. B., & Forster, R. E. (1966). The effect of size of red cells on the kinetics of their oxygen uptake. *Journal of General Physiology, 49*, 727–742.

Hooke, R. (2003). *(1665) Micrographia; or some physiological descriptions of minute bodies made by magnifying glasses with observations and inquiries thereupon.* Dover.

Howard, J. M., Griffis, H.-B., Westendorf, R., & Williams, J. B. (2020). The influence of size and abiotic factors on cutaneous water loss. *Advances in Physiology Education, 44*, 387–393.

Jones, D. A. (1979). The importance of surface area/volume ratio to the rate of oxygen uptake by red cells. *Journal of General Physiology, 74*, 643–646.

Kattge, J., Bönisch, G., Díaz, S., et al. (2020). TRY plant trait database—Enhanced coverage and open access. *Global Change Biology, 26*, 119–188. https://doi.org/10.1111/gcb.14904

Martins, E., Rapp, H. T., Xavier, J. R., Diogo, G. S., Resi, R. L. & Silva, T. H. (2021). Macro and microstructural characteristics of North Atlantic deep-sea sponges as bioinspired models for tissue engineering scaffolding. *Frontiers in Marine Science, 7,* 18 pp. https://doi.org/10.3389/fmars.2020.613647

Meeh, K. (1879). Oberflächenmessungen des menschlichen Körpers. *Zeitschrift fur biologie, 15,* 425–458.

Newman, M. T. (1953). The application of ecological rules to the racial anthropology of the aboriginal New World. *American Anthropologist, 55,* 311–327.

Nguyen, V., Lilly, B., & Castro, C. (2014). The exoskeletal structure and tensile loading behavior of an ant neck joint. *Journal of Biomechanics, 47,* 497–504.

Nudds, R. L., & Oswald, S. A. (2007). An interspecific test of Allen's Rule: Evolutionary implications for endothermic species. *Evolution, 61,* 2839–2848. https://doi.org/10.1111/j.1558-5646.2007.00242.x

Santavy, D. L., Courtney, L. A., Fisher, W. S., Quarles, R. L., & Jordan, S. J. (2013). Estimating surface area of sponges and gorgonians as indicators of habitat availability on Caribbean coral reefs. *Hydrobiologia, 707,* 1–16.

Schmidt-Nielsen, K. (1984). *Scaling: Why is animal size so important?* Cambridge University Press, 241 pp.

Schultheiss, P., Nooten, S. S., Wang, R., Wong, M. K. L., Brassard, F., & Guénard, B. (2022). The abundance, biomass, and distribution of ants on Earth. *Proceedings National Academy of Sciences,* 119. https://doi.org/10.1073/pnas.2201550119

Sereno, M. I., Diedrichsen, J., Tachrount, M., Testa-Silva, G., d'Arceuil, H., & De Zeeuw, C. (2020). The human cerebellum has almost 80% of the surface area of the neocortex. *Proceedings of the National Academy of Sciences,* 117. https://doi.org/10.1073/pnas.2002896117

Sender R., Fuchs S., & Milo R. (2016) Are we really vastly outnumbered? Revisiting the ratio of bacterial to host cells in humans. *Cell, 164,* 337–340.

Wojtusiak, J., Godzińska, E. J., & Dejean, A. (1995). Capture and retrieval of very large prey by workers of the African weaver ant, *Oecophylla longinoda* (Latreille 1802). *Tropical Zoology, 8,* 309–318. https://doi.org/10.1080/03946975.1995.10539287

Zafren, K. (2013). Frostbite: Prevention and initial management. *High Altitude Medicine & Biology, 14,* 9–12.

Chapter 6
Biochemistry

6.1 Protein Folding

A protein is a macromolecule consisting of long chains of amino-acid residues; these chains are known as polypeptides. Each amino acid consists of an amine group (H_2N), a carboxylic acid group (COOH), a hydrogen atom, and a residue or side chain (an R group), all linked to a central carbon atom (or alpha carbon; C_α) (Fig. 6.1, conveniently reproduced from Fig. 5.1). The side chains (or residues) contain C, H, and O; some also contain N and/or S. Every R group has a distinctive chemical composition, size, shape, electric charge, and reactivity. Amino acids are ordered into categories (e.g., non-polar and neutral; polar and neutral; polar and acidic; polar and basic) based on the properties of the side chains.

Although more than 500 amino acids exist in nature, the human genome contains only 20. These 20 include: (1) so-called "non-essential" amino acids that the human body can manufacture (e.g., alanine, cysteine, glutamic acid, glutamine, glycine, and proline) and (2) "essential" amino acids that must be obtained from food (e.g., isoleucine, lysine, methionine, phenylalanine, tryptophan, and valine). Amino acids have various properties including size, electric charge, hydrophilicity (attraction to water molecules) and hydrophobicity (repulsion by water molecules). These properties determine the positions of the amino acids within the protein as well as the overall protein structure. For example, water-soluble proteins tend to have hydrophilic residues on their surface, exposed to the surrounding aqueous fluid. In contrast, hydrophobic residues tend to be buried in the middle of the protein.

Proteins are distinguishable by their particular sequences of amino-acid residues. These sequences are responsible for protein folding into distinct three-dimensional structures (e.g., folds, curves, coils) that govern protein biological activity. Chemical bonds among the components of the chains preserve the protein shapes.

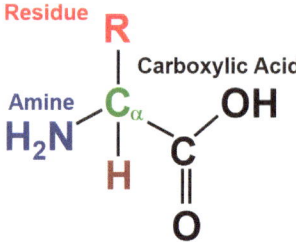

Fig. 6.1 Structure of an amino acid. Every amino acid consists of an amine group (in blue), a carboxylic acid group (black), a hydrogen atom (brown), and a residue or side chain (red R). They are all connected to a central C atom (or α-carbon) (green), designated C_α. Each amino acid has a unique residue containing C, H, and O; some also contain N and/or S

There are three common protein types:

(1) Globular proteins (a.k.a. spheroproteins) tend to be fairly compact and somewhat water-soluble. They have several different functions including storage (e.g., ferritin), hormones (e.g., insulin), antibodies, enzymes, and transport (e.g., hemoglobin).

(2) Fibrous proteins (a.k.a. scleroproteins) typically have low water-solubility. They consist of elongated, fibrous polypeptide chains; many serve as structural components (e.g., collagen) or are involved in muscle contraction (e.g., myosin).

(3) Membrane proteins can serve as receptors or are conduits for molecular transport through cell membranes. Most have alpha-helices. It seems likely that most proteins in the human body interact with membranes.

How spherical are the globular proteins? As shown earlier, a perfect sphere has a surface area proportional to the two-thirds power of its volume: $SA \propto V^{2/3}$. Expressed in decimal notation, this can be written as $SA \propto V^{0.67}$. For globular proteins, the surface area is proportional to the molecular weight of the protein raised to the power of 0.73 (Miller et al., 1987). Because molecular weight is a three-dimensional parameter related to volume, the relationship of surface area to volume follows a similar power law. However, examination of the three-dimensional structures of globular proteins shows them generally to be irregular clumps, impossible to confuse with spheres. The protein database website, www.rcsb.org, contains every known protein structure, viewable in three dimensions.

Recently developed artificial-intelligence tools can design novel proteins with specific structures by finding amino-acid sequences that fold into predetermined shapes (Dauparas et al., 2022; Wang et al., 2022; Wicky et al., 2022). The tools can also conduct random searches of possible protein sequences to select those with specific capabilities. The hope is that these new proteins will be able to perform important functions such as unclogging arteries, combatting novel viruses, and digesting tiny plastic particles in the wild.

Proteins bend themselves into four basic structural types – primary, secondary, tertiary, and quaternary – showing increasing three-dimensional complexity. The final structure that a protein folds itself into is the thermodynamically most favored state.

Primary structure: This is just the simple polypeptide chain made of the particular amino-acid residues characteristic of that protein. Insulin, for example, is a hormone produced in the pancreas; it has two distinct polypeptide chains (one with 21 amino-acids residues, the other with 30) linked together with two disulfide bonds.

Secondary structure: Atoms of the polypeptide chain interact and produce local folding. There are two common folding forms: (1) The *alpha helix* is a right-handed helix in which every N–H amino group undergoes hydrogen bonding with the C==O carbonyl group. Common amino acids that frequently occur with this shape are alanine, leucine, and methionine. (2) The *beta pleated sheet* consists of polypeptide sheets connected laterally by hydrogen bonds. Amino acids that often display this structure include threonine, valine, and isoleucine.

Tertiary structure: These are the full three-dimensional structures of proteins, enabling them to become functional. Their characteristic shapes are due to bond interactions between the side chains (R groups) of the amino acids in the polypeptide chains (which form the "backbone" of the structure). An example of a tertiary structure is myoglobin (Fig. 6.2), a protein found in cardiac and muscle cells. It consists of 153 amino acids in the shape of eight alpha helices connected to each other by loops.

Quaternary structure: Whereas most proteins consist of single polypeptide chains exhibiting primary, secondary, and tertiary structures, a few proteins contain multiple

Fig. 6.2 Three-dimensional structure of the myoglobin protein. Alpha-helices are shown in blue. A heme group (gray) with a bound oxygen molecule (red) occurs among the coils at center right. *Image* by Aza Toth

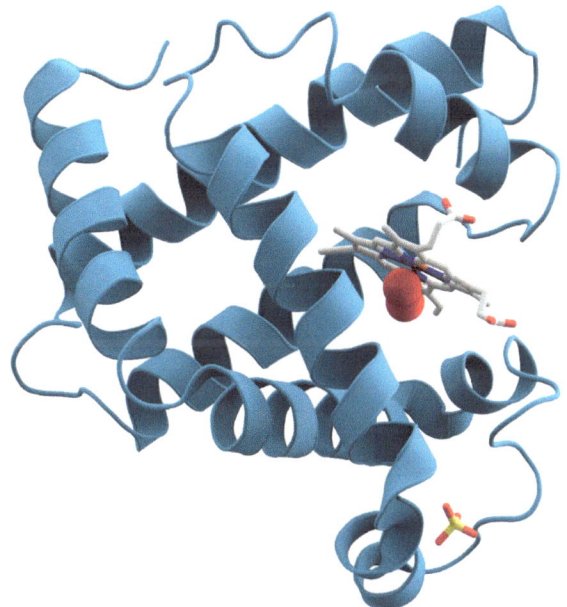

Fig. 6.3 Structure of hemoglobin. Alpha-helices (red), beta-sheets (blue) and iron-bearing heme groups (green) are shown. *Image* from Zephyris at English Wikipedia

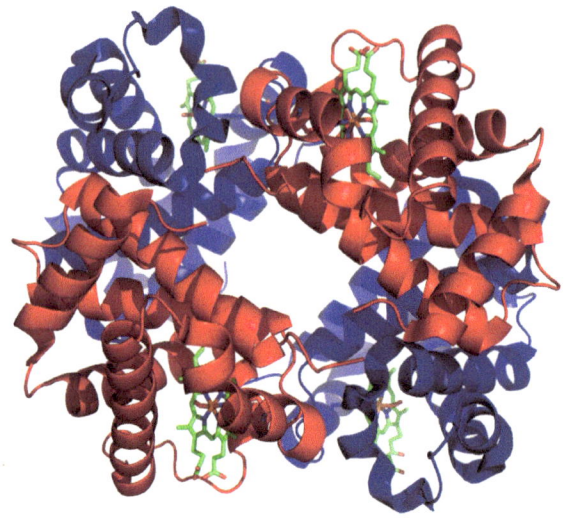

polypeptide chains linked together to form a quaternary structure. One example is hemoglobin (Fig. 6.3), an iron-rich, oxygen-transporting protein that makes up about 96% of the dry weight of red blood cells. Most of the amino acids in hemoglobin are in alpha-helix shapes stabilized by hydrogen bonds. They are joined together by short non-helical structures.

A summary of the increasing complexity shown in protein folding appears in Fig. 6.4. The driving forces that stabilize the tertiary and quaternary structures are hydrogen bonding, electrostatic interactions, hydrophobicity, and disulfide bonding.

The deep interior of a protein, called the protein core, is the portion of a folded protein inaccessible to the aqueous solvent surrounding the protein. The accessible surface area of a protein is the area at the surface of a protein that is accessible to the surrounding solvent. It is usually calculated as the contact area that hypothetically would be traced out by a small sphere of a particular radius rolled over the outside of the protein. Many proteins (called mosaic proteins) have distinct regions that have folded themselves into different shapes independently of the rest of the protein. These distinct regions are called protein domains (e.g., Fig. 6.5).

Larger proteins have smaller surface/volume ratios than smaller proteins of the same shape. Because protein surfaces tend to be the site of hydrophilic amino acid residues, decreases in the surface/volume ratio that occur with increasing protein size should go along with a higher proportion of hydrophilic amino-acid residues occurring in the protein core.

Shirota et al. (2008) defined a particular surface/volume ratio (SVR) as the ratio between accessible surface area and the volume of the domain of a protein. The measured values of this ratio range from 0.19 to 0.45 Å^{-1} (where 1 Å is one angstrom, equivalent to 10^{-10} m or 0.1 nm). They found that, with increasing SVR among domains in different proteins, higher proportions of hydrophilic amino-acid residues (e.g., glutamic acid, glutamine, arginine) were buried in the protein core and

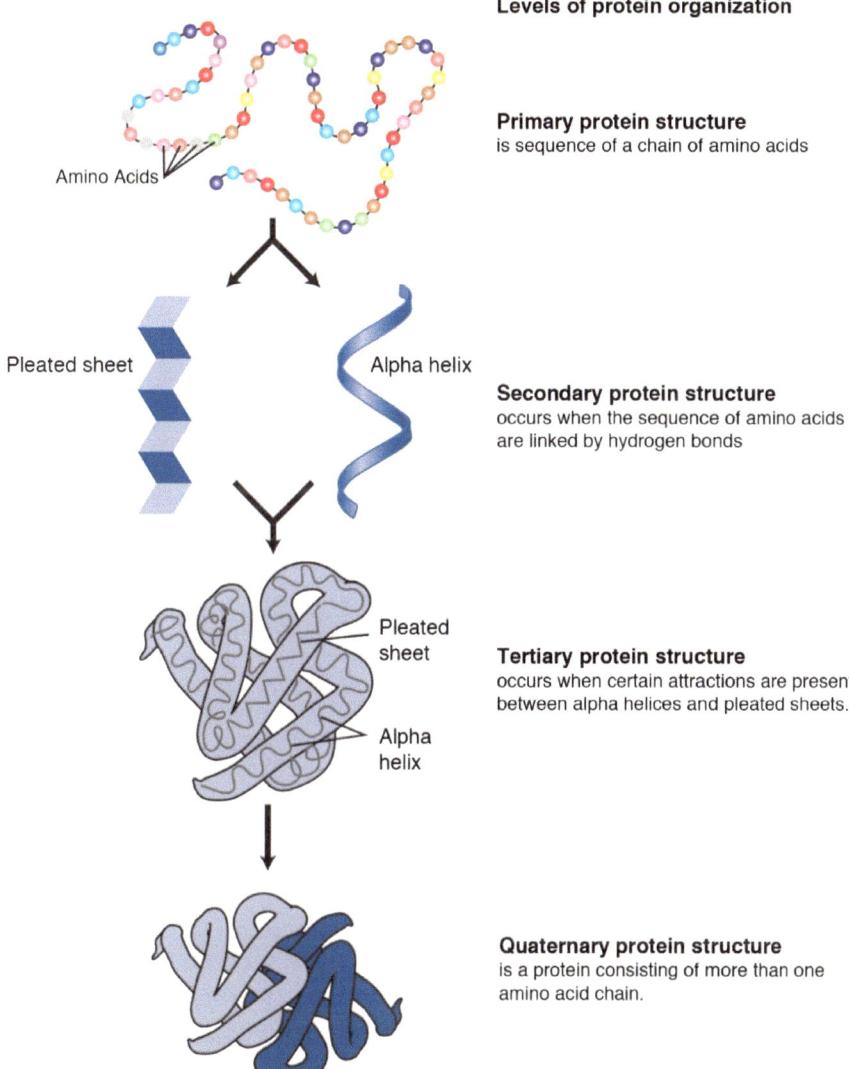

Levels of protein organization

Amino Acids

Primary protein structure
is sequence of a chain of amino acids

Pleated sheet

Alpha helix

Secondary protein structure
occurs when the sequence of amino acids
are linked by hydrogen bonds

Pleated
sheet

Tertiary protein structure
occurs when certain attractions are present
between alpha helices and pleated sheets.

Alpha
helix

Quaternary protein structure
is a protein consisting of more than one
amino acid chain.

Fig. 6.4 Levels of protein organization. *Image* from the National Human Genome Research Institute

higher proportions of hydrophobic amino-acid residues (e.g., tyrosine, tryptophan, phenylalanine) appeared at the domain surface. The proteins evolved these particular configurations to optimize their function. It seems plausible that the relatively high hydrophobicity of the surfaces of these proteins enables them to coexist in the densely packed environments inside cells.

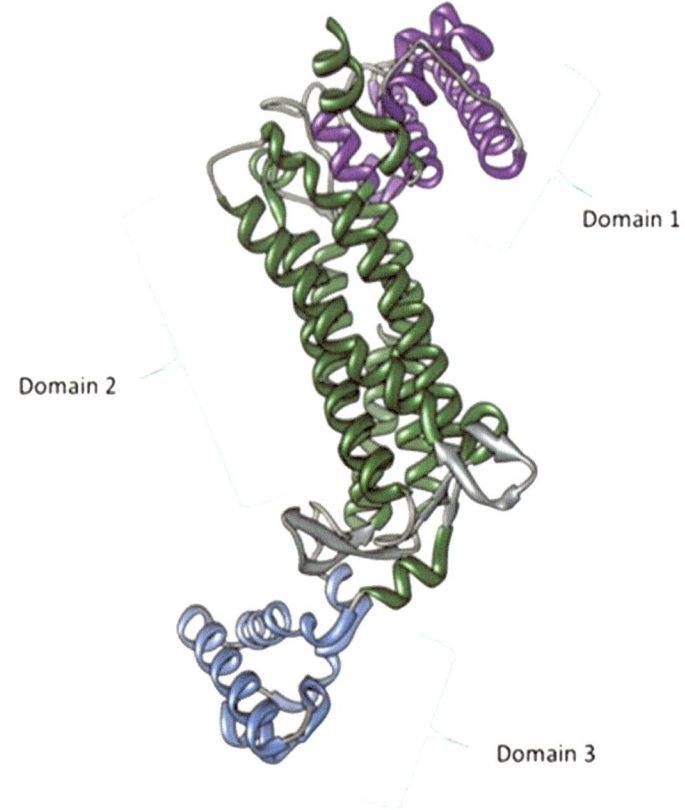

Domain 1

Domain 2

Domain 3

Fig. 6.5 A protein consisting of three independently folded domains. *Image* modified from SDFayson

6.2   A Summing Up

Consequences of the surface/volume ratio of three-dimensional structures permeate the living world (Fig. 6.6). In the human body, they define the internal shapes of major organs including the brain, lungs, and small intestine. The surface/volume ratio influences the properties of proteins in every cell. As declared by the Centre for Geometric Biology of Monash University (Melbourne, Australia): "Geometric biology allows us to understand the dynamics of how living things convert energy flows into mass, at all scales of biological organisation. The size and shape (together, the 'geometry') of organisms ultimately determine these flows."

Fig. 6.6 Shell of a cephalopod of the species *Nautilus macromphalus*, demonstrating a logarithmic spiral. *Modified image* from User: Mgiganteus1

References

Dauparas, J., Anishchenko, I., Bennett, N., Bai, H., Ragotte, R. J., Milles, L. F., Wicky, B. I. M., Courbet, A., De Haas, R. J., Bethel, N., Leung, P. J. Y., Huddy, T. F., Pellock, S., Tischer, D., Chan, F., Koepnick, B., Nguyen, H., Kang, A., Sankaran, B., Bera, A. K., King, N. P., & Baker, D. (2022). Robust deep learning-based protein sequence design using ProteinMPNN. *Science, 378*, 49–56. https://doi.org/10.11126/science.add2187

Miller, S., Janin, J., Lesk, A. M., & Chothia, C. (1987). Interior and surface of monomeric proteins. *Journal of Molecular Biology, 196*, 641–656.

Shirota, M., Ishida, T., & Kinoshita, K. (2008). Effects of surface-to-volume ratio of proteins on hydrophilic residues: Decrease in occurrence and increase in buried fraction. *Protein Science, 17*, 1596–1603.

Wang, J., Lisanza, S., Juergens, D., Tischer, D., Watson, J. L., Castro, K. M., Ragotte, R., Saragoui, A., Milles, L. F., Baek, M., Anishchenko, I., Yang, W., Hicks, D. R., Expòsit, M., Schlichthaerle, T., Chun, J.-H., Dauparas, J., Bennett, N., Wicky, B. I. M., Muenks, A., Dinaio, F., Correia, B., Ovchinnikov, S., & Baker, D. (2022). Scaffolding protein functional sites using deep learning. *Science, 377*, 387–394.

Wicky, B. I. M., Milles, L. F., Courbet, A., Ragette, R. J., Dauparas, J., Kinfu, E., Tipps, S., Kibbler, R. D., Baek, M., Dimaio, F., Li, X., Carter, L., Kang, A., Nguyen, H., Bera, A. K., & Baker, D. (2022). Hallucinating symmetric protein assemblies. *Science, 378*, 56–61. https://doi.org/10.1126/science.add1964

Chapter 7
Chemical Reactions

7.1 Elementary Chemistry

Atomic nuclei consist of protons (each with a positive electric charge of $+1$) and neutrons (neutral particles with no charge). Surrounding the nucleus are clouds of electrons; each electron has a negative electric charge of -1. The charges of the proton and electron are of identical strength despite the fact that an electron's mass is only 0.05447% that of a proton. Atoms are electrically neutral: the number of protons equals the number of electrons.

The number of protons is called the atomic number, symbolized by the letter "Z", from the German word for number: *Zahl*. The atomic number defines the element—if an atom has six protons, it is carbon; if it has 26 protons, it is iron; if it has 92 protons, it is uranium. Every atom except common hydrogen (whose nucleus consists of a single proton) contains one or more neutrons. Because the mass of a neutron is very close to that of a proton (the neutron being only about 0.1% heavier), the atomic weight of an atom is the sum of the number of protons and the number of neutrons in its nucleus. Beryllium, for example, has four protons (giving it an atomic number of 4) and five neutrons (giving it an atomic weight of 9) (Fig. 7.1).

Isotopes are elements with the same atomic number (i.e., the same number of protons), but with different numbers of neutrons. If an isotope is not radioactive, it is said to be stable; its nucleus will remain intact over time. Oxygen, for example, has three stable isotopes—all have eight protons (making it oxygen), but the number of neutrons can be eight (yielding an atomic weight of 16: (^{16}O), nine (atomic weight of 17: ^{17}O), or ten (atomic weight of 18: ^{18}O). These isotopes are not equally abundant: 99.74% of all oxygen atoms are ^{16}O, 0.0367% are ^{17}O, and 0.187% are ^{18}O. Oxygen also has many radioactive isotopes with half-lives ranging from 1.98×10^{-22}s (^{11}O with three neutrons) to 122.24s (^{15}O with seven neutrons). ^{13}O decays to ^{13}N by electron capture; ^{15}O decays to ^{15}N and emits a positron. The heaviest oxygen isotope is ^{28}O (harboring 20 neutrons); it is also radioactive with a half-life less than one ten-millionth of a second. [A half-life is the time it takes for 50% of the atoms in a given batch of the isotope to decay.]

© The Author(s), under exclusive license to Springer Nature Switzerland AG 2023
A. E. Rubin, *Surface/Volume*, https://doi.org/10.1007/978-3-031-23749-2_7

Fig. 7.1 A simplified structure of beryllium 9. There are four protons and five neutrons in the nucleus. The innermost, lowest-energy subshell of electrons (the 1s subshell) is closest to the nucleus; it contains two electrons. The outer electron subshell (2s) is of higher energy; it also contains two electrons

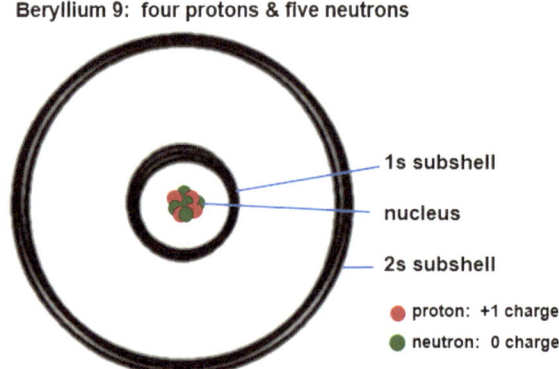

Beryllium 9: four protons & five neutrons

1s subshell

nucleus

2s subshell

● proton: +1 charge
● neutron: 0 charge

Electrons occur in clouds around the nucleus. These clouds are actually probability distributions in which individual electrons roam; they are constantly in motion. There is a 90% chance that a particular electron will be in its designated cloud at any given moment. The lowest-energy cloud (also called a subshell or an orbital) is designated 1s; it is closest to the atomic nucleus. The 2s cloud is located farther from the nucleus. The 1s and 2s clouds are spheres. The 2p orbital has the same energy as 2s but is shaped like a dumbbell. Higher-energy clouds are more distant from the nucleus; the 3d orbital is shaped like a set of two perpendicular dumbbells; the 4f orbital is complex with four distinct shapes (Fig. 7.2).

Due to quantum–mechanical restrictions, there is a limit on the number of electrons that can occupy a single orbital. The 1s and 2s orbitals can each hold up to two electrons; the 2p orbitals can accommodate up to six electrons; the 3d orbitals can house up to ten; the 4f orbitals, up to 14. There are seven orbitals in the 4f subshell; each orbital can accommodate two electrons with opposite spin.

As elements increase in atomic number, the numbers of electrons in clouds around the nucleus increase in lockstep to maintain atomic electrical neutrality. Once an electron orbital is filled, additional electrons start to fill new orbitals.

To a large extent, the number of electrons in the outermost electron shell (i.e., the valence electrons) determines the chemical properties of an element; this is because it is these electrons that are shared or exchanged with those of other atoms during bonding. The noble gases (occurring in the right-most column of the periodic table; Fig. 7.3) contain completely filled electron shells. They are thus chemically inert; they form very few compounds with other elements.

Ions are elements that have gained or lost one or more electrons from their outermost shells. They are not electrically neutral. Atoms that have *lost* one or more electrons (cations) have positive charges; atoms that have *gained* one or more electrons (anions) have negative charges.

The periodic table (Fig. 7.3) is a graphical arrangement of all known elements. Elements with similar electron configurations (i.e., those with the same number of electrons in their outermost shells) are aligned in the same column; heavier elements (containing more protons and neutrons) are placed beneath lighter ones.

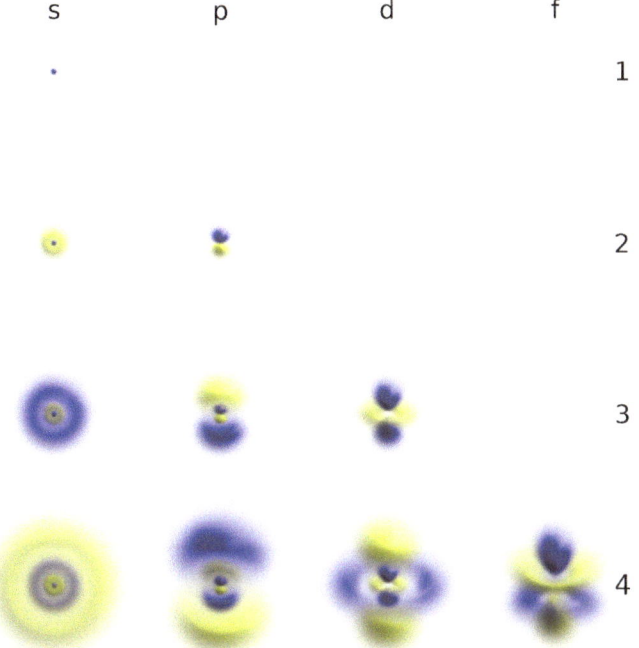

Fig. 7.2 Electron shells showing the 1s, 2s, 2p, 3s, 3p, 3d, 4s, 4p, 4d, and 4f orbitals. The letters above each column designate the angular momentum quantum number (l), describing the shape of the orbital. The numbers to the right of each row denote the principal quantum number (n), describing the energy state. Every element that contains electrons in 4f orbitals, also has electrons in 5s and 5p orbitals. Image by Geek3

Electron shells that are full or half-full are stable, low-energy configurations. That is why noble gases, with their filled shells, tend to be unreactive.

There are three principal types of chemical bonds that involve the gain, loss, or sharing of valence electrons: ionic, covalent, and metallic.

Ionic bonding involves an electrostatic attraction between cations and anions caused by the cation donating one or more electrons from its outer shell and the anion acquiring those electrons in *its* outer shell. Each ion develops the stable configuration of a filled electron shell. An example of a substance with ionic bonding is halite (common table salt), NaCl. Covalent bonding occurs when the electron orbitals between adjacent atoms overlap. Electrons are attracted to the positively charged nuclei of both atoms; this produces an electrostatic bond because both atoms share the orbiting electrons. Diamond, made of covalently bonded carbon atoms, appears to be the hardest, naturally occurring material. Metallic bonding, as the name implies, is present in metals: metal cations are compacted into a crystal lattice; the valence electrons are no longer restricted to shells around a single atom and move freely throughout the structure. Silver and copper are materials with metallic bonding.

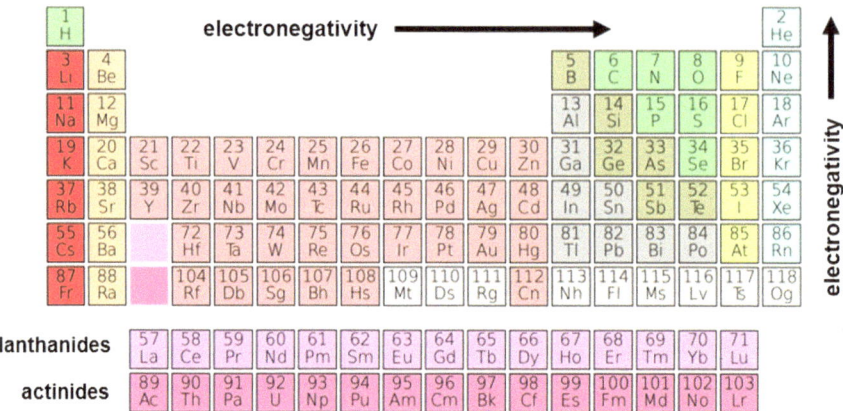

Fig. 7.3 The periodic table of the elements, showing the one-or-two-letter element symbols and the atomic numbers (equivalent to the number of protons in the nucleus). After the right-most column of noble gases is excluded, there is a strong trend for electronegativity to increase from left to right with increasing numbers of protons in the atomic nucleus. Electronegativity also increases from the bottom toward the top (reflecting ever fewer electron shells shielding the nucleus from the outer electrons). Fluorine has the greatest electronegativity value; oxygen has the second highest

Electronegativity is a chemical property that quantifies the tendency of an atom to attract a bonding pair of electrons. Remember: opposite charges attract; like charges repel. Atoms with higher atomic numbers (i.e., with greater numbers of protons) exert a stronger pull on their outermost electrons and thus tend to have higher electronegativity values. That is why electronegativity increases from left to right across the columns of the periodic table along with the number of protons in the nucleus. (This increase in electronegativity does not include the non-reactive noble gases in the rightmost column.)

Heavier elements have more protons in the nucleus, so there is greater positive charge available to attract electrons. However, for elements in the lower rows of the periodic table, the outermost electrons are shielded from the nucleus by inner electron shells, diminishing the electrostatic attraction between the protons in the nucleus and the valence electrons. Consequently, electronegativity increases from the bottom of the table toward the top (as well as from left to right). Thus, elements with the highest electronegativity values are located at the upper right of the periodic table. Fluorine has the highest electronegativity; oxygen has the second highest (Fig. 7.4). As an illustration, in the compound hydrogen fluoride (HF), the H atom has an electronegativity of 2.20 and the F has an electronegativity of 3.98; thus, the bonding electrons will be closer to the F atom than the H atom.

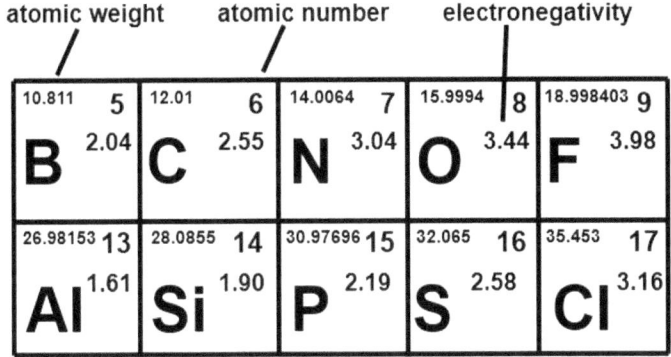

Fig. 7.4 Upper-right portion of the periodic table showing atomic weight, atomic number and electronegativity value (Pauling scale). Electronegativity increases from left to right and from bottom to top. The right-most column of noble gases has been omitted

7.2 Chemical Properties of Water

Water has been called the "universal solvent" because it can dissolve more substances than any other liquid. To understand water's unique properties, we need to check its chemical formula (H_2O), consult the periodic table, and explore the implications of the electronic configuration (the distribution of electrons) of the constituent atoms.

Oxygen has eight electrons. Two electrons fill its 1s subshell; they are unavailable for forming chemical bonds. In the water molecule, the six electrons in the second electron shell around the oxygen nucleus do not reside in 2s and 2p orbitals, but instead are housed in hybrid (so-called sp^3) orbitals. The oxygen forms a very strong covalent bond (called a sigma bond) with each hydrogen atom: hydrogen shares its sole electron with the oxygen atom, effectively filling the 1s subshell of each hydrogen (Fig. 7.5). That leaves the oxygen atom with two pairs of unbonded (unshared) electrons in hybrid orbitals in its 2nd-level subshell. These two sets of unbonded electrons tend to repel one another; the most stable configuration occurs when these lone pairs are farthest apart.

The unshared electron pairs around the oxygen atom are a little more electrically repulsive than the bonded electrons, bending the water molecule into a quasi-tetrahedral structure (Fig. 7.6). Oxygen is at the center. Two of the four vertices are occupied by hydrogen atoms; the other two are occupied by the two lone pairs of unshared electrons around the oxygen. Both hydrogen atoms are displaced to one side of the molecule, away from the oxygen. However, the hydrogen atoms cannot get too close together because the positive charges of their nuclei repel each other. The resulting electrostatic equilibrium distorts the tetrahedron: instead of the 109.47° angle of a regular tetrahedron, the H–O–H bend angle (in the gas phase) is 104.48°.

Due to oxygen's high electronegativity (3.44) relative to hydrogen (2.20), the electrons that form covalent bonds with hydrogen spend more time near the oxygen atom than the two hydrogen atoms. This results in the region around the oxygen

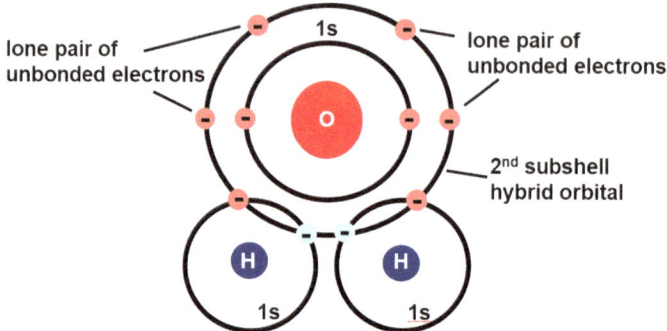

Fig. 7.5 Covalent bonding in the water molecule. H = hydrogen nucleus (blue); O = oxygen nucleus (red); circles with minus signs are electrons. The two H atoms share their two electrons (light blue) with two of the electrons from oxygen (light red). This leaves oxygen with two pairs of unbonded electrons in its 2nd subshell

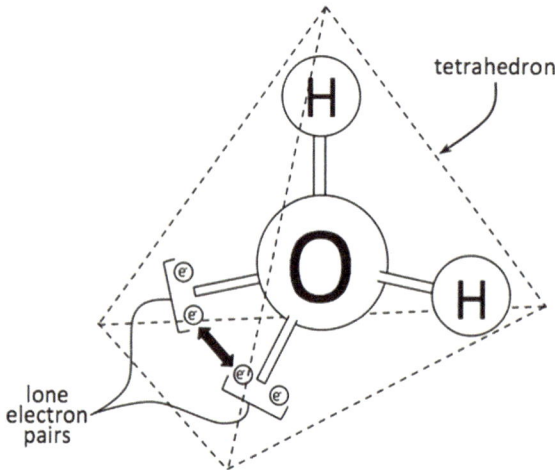

Fig. 7.6 The quasi-tetrahedral structure of the water molecule. Two of the vertices are occupied by H atoms; two are occupied by the pairs of unshared electrons. Diagram modified from one by Thebiologyprimer

atom in the bent water molecule having a net partial negative charge ($\delta-$) while the region around the hydrogen atoms has a net partial positive charge ($\delta+$) (Fig. 7.7). A dipole is thus created—a hypothetical line pointing from the space between the two hydrogen atoms toward the oxygen atom. The water molecule is therefore described as "polar". Such molecules tend to line up in the presence of an electric field.

Because individual water molecules have a net positive charge on one side (near the hydrogen atoms) and a net negative charge on the other (near the oxygen atom), they produce an electrostatic attraction to neighboring water molecules. The oxygen side of one molecule tends to bond with the hydrogen side of another. This is a hydrogen

Fig. 7.7 The polar water molecule. The numerical values show the electronegativities. The two pairs of unbound electrons are shown at the top. The H–O–H angle is 104.48°. Diagram modified from one by Riccardo Rovinetti

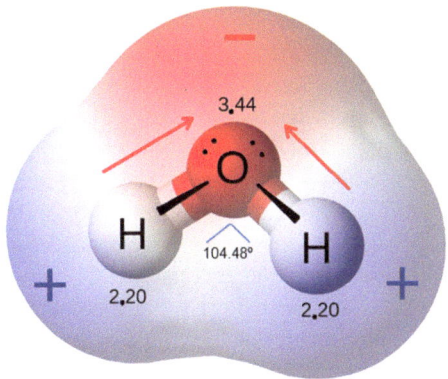

bond; although weaker than a covalent bond, it is the strongest intermolecular binding force. A fair amount of energy is required to break hydrogen bonds—this is the reason why water exhibits its unusual collective properties: its relatively high values for the melting point, boiling point, heat of fusion (the amount of energy required to transform a specific amount of a solid into a liquid without a change in temperature), heat of vaporization (the amount of energy necessary to transform a specific amount of liquid into a vapor at constant temperature and pressure), and surface tension (the ability of a liquid to resist an external force).

The hydrogen bonds also account for normal ice (Ice 1 h) being less dense than liquid water. In warm water, the molecules move past each other rapidly and hydrogen bonds can break and reform. As the water cools, the kinetic energy decreases and fewer hydrogen bonds break apart. During crystallization, the water molecules are locked into a low-density structure with hexagonal closest packing (HCP) with 26% void space.

The hydrogen bonding of liquid water also causes it to exhibit appreciable cohesion. A large proportion of the water molecules in a given spot is held together by these hydrogen bonds. Adhesion is another physical property caused by water's polar nature; water may form a thin film on a smooth surface such as glass (if the glass has a slight charge) because the electrostatic attraction of the water molecules for the glass exceeds the cohesive forces among the water molecules. Capillary action is caused by the water molecules adhering to the sides of a narrow tube pulling adjacent water molecules along with them up the tube via cohesion. In many cases, this is opposite to the direction of gravity (as in the xylem vascular tissues in a tall tree).

7.3 Dissolution

Dissolution (also known as solvation) is the process wherein one or more substances (i.e., the solute(s)) form a homogeneous chemical solution with a more abundant substance (the solvent). Solvents can be gases, liquids, or solids. After mixing, the solution has the same physical state (i.e., solid, liquid, or gas) as the solvent. The most common chemical solution present at the Earth's surface is air—a fairly homogeneous mixture consisting of a solvent (molecular nitrogen, N_2; 78%) and several solutes: molecular oxygen, O_2 (21%); water vapor, H_2O (0–3%); argon, Ar (0.93%); carbon dioxide, CO_2 (0.04%); and trace amounts of neon (Ne), helium (He), methane (CH_4), and krypton (Kr). The most abundant liquid solvent, of course, is water. Water's dissolution properties are best illustrated with salt and sugar.

Salt (NaCl) is an ionically bonded compound; when dissolved, particles of salt dissociate into Na^+ cations and Cl^- anions. These individual ions are surrounded by a hydration shell of water molecules in which the oppositely charged sides of the water molecules face the dissolved ion. The oxygen side of the water molecules (with its partial negative charge) faces the positively charged sodium cation (Fig. 7.8). Where a Cl^- anion occurs, the water molecules face the other direction. Although the ionic bond in NaCl is strong, the combined electrostatic attraction of the numerous surrounding water molecules is sufficient to break the bond and separate the sodium and chlorine ions. A homogeneous solution is formed. Ionic compounds dissolved in water can conduct electricity and are known as electrolytes.

Table sugar (or sucrose, $C_{12}H_{22}O_{11}$) is composed of two simpler sugars: one molecule of glucose ($C_6H_{12}O_6$ forming a six-membered carbon ring structure) attached to one molecule of fructose ($C_6H_{12}O_6$ forming a five-membered carbon ring structure). Individual sucrose molecules are held together by strong covalent bonds (called glycosidic bonds). There are many polar O–H bonds in sucrose resulting in different portions of the molecule having a partial positive charge (near the hydrogen atoms) and a partial negative charge (near the oxygen atoms). Separate sucrose molecules are held together by the (much-weaker) hydrogen bonds between the oppositely charged polar areas.

When sugar dissolves in water, the strong covalent bonds within individual sugar molecules keep individual molecules intact. Nevertheless, the combined electrostatic

Fig. 7.8 A dissolved sodium cation, derived from table salt, is surrounded by a hydration shell of water molecules. The delta symbol (δ) represents a partial charge: positive around the hydrogen atoms, negative around the oxygen. Diagram by Taxman

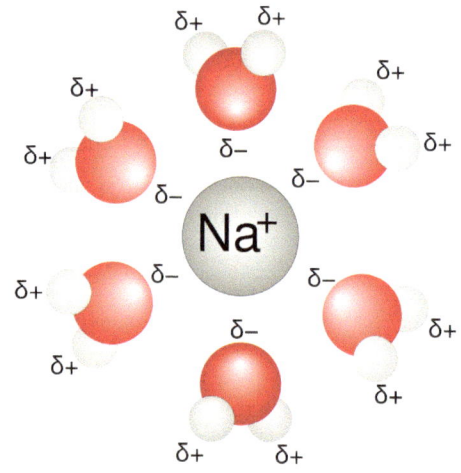

attraction of the water molecules can overcome the intermolecular forces (i.e., the hydrogen bonds) in the sugar, ripping one sugar molecule away from another. A sugary solution is composed of water molecules forming hydration shells around individual intact sugar molecules. The positively charged hydrogen sides of the water molecules face the negatively charged oxygen atoms in the sugar. The solution is chemically homogeneous, but because the covalently bonded solute has not broken apart into individual ions, it is not electrically conductive.

There is a limit on how much solute can dissolve in a solvent at a given temperature and pressure. For example, the maximum solubility of salt (NaCl) in water at 25 °C is 357 mg of salt per milliliter of water. This limit is called the saturation point. If more salt is added or if some water is removed, salt crystals will precipitate from the water and settle to the bottom of the container. Settling occurs because the density of salt crystals (halite; 2.16 g cm^{-3}) is much higher than that of liquid water (1.0 g cm^{-3}).

Rock salt deposits around the world formed after evaporation of seawater increased the concentration of salt beyond the saturation point. Beneath the floor of the Mediterranean Sea is a 1.5-km-thick layer of salt (dubbed the Mediterranean Salt Giant). It was deposited between 5.96 and 5.33 million years ago when the Strait of Gibraltar was closed by tectonic forces in the Earth's crust, a closure that prevented Atlantic-Ocean water from replacing water evaporated from the Mediterranean Sea. The level of the sea gradually dropped by more than a kilometer and the dissolved salt concentration increased beyond the saturation point. Salt crystals precipitated from the water and settled to the bottom. Around 5.6 million years ago, the sea may have evaporated almost completely. This period of desiccation is known as the Messinian Salinity Crisis. When the strait reopened 5.33 million years ago (e.g., Blanc, 2002), water from the Atlantic poured in: discharge rates may have exceeded 100 million cubic meters per second for several years in an hypothesized event called the Zanclean Flood.

7.4 The Effort to Reduce Sodium in Food

Table salt is composed of 39.34 wt.% sodium and 60.66 wt.% chlorine; it provides about 90% of the sodium in the human diet. Sodium helps maintain cell membrane potential (allowing ions to be distributed properly between the inside and outside of a cell) and facilitates the optimal absorption of nutrients in digested food by the small intestine. Retention of sodium in the body is accompanied by retention of water; hence, sodium content controls blood volume and, as a consequence, blood pressure. [Normal blood pressure for an adult should be between 90/60 and 120/80 mm Hg (millimeters of mercury, where 1 mm Hg is approximately 1 torr). The first number (systolic pressure) measures the force blood exerts on arterial walls while the heart is pumping. The second number (diastolic pressure) measures the force on arterial walls between heart beats.]

Processed foods provide 75–80% of the daily sodium intake. In descending order, the top five contributors of sodium in the typical U.S. diet are (1) bread and rolls, (2) mammalian meat, (3) pizza, (4) poultry, and (5) soup (Mueller et al., 2016). There are advantages for foods to contain sodium; it imparts a widely appreciated clean, salty taste, it enhances the flavor of other aromatic compounds (fragrant chemicals with stable hydrocarbon ring structures), it decreases bitterness, and it inhibits the growth of pathogens (thereby increasing shelf life) by decreasing water activity. Sodium chloride elicits a salty taste only when it is dissolved in water (or saliva), enabling its detection by taste receptor cells.

But too much sodium in the body can be damaging. The World Health Organization (WHO) recommends a daily intake for adults of less than 5 g of salt (equivalent to less than 2 g of sodium), but in Western Europe, the daily sodium intake is 3.8 g, and in Central Asia, 5.5 g. More than 88% of adults across the globe ingest at least 3 g of sodium a day. These elevated sodium levels are associated with hypertension (blood pressures exceeding about 140/90 mm Hg). Hypertension, in turn, is responsible for 62% of strokes and 49% of cases of coronary heart disease (He & MacGregor, 2008).

In 2015, the National Association of Pizza Operators (NAPO) reported that Americans eat 3 billion pizzas a year, equivalent to 46 slices for every man, woman, and child in the country. Because pizza is among the most significant sources of sodium in the American diet, taste tests have been used to evaluate strategies for decreasing the amount of sodium in pizza crust. Successful strategies include an immediate salt reduction of 10% (greater reductions are noticeable) and the substitution of 30% potassium chloride (sylvite, KCl) for NaCl (greater substitutions result in a bitter, metallic taste).

Tests show that salty taste can be enhanced by accelerating the release of sodium, particularly during the first 15s of chewing. This leads to alternative strategies for reducing sodium by changing the surface/volume ratio of salt crystals. For example, SODA-LO® Salt Microspheres, manufactured by the British-based company Tate & Lyle PLC, are small, hollow, crystalline NaCl microspheres. Compared to normal cubic halite crystals, the microspheres have very high surface/volume ratios. A 2013 company press release claims use of SODA-LO® can reduce the overall amount of salt in baked goods and snacks by 25–50% without affecting the salty taste. Other salt shapes with high surface/volume ratios include: (1) long, flat, low-density flakes (added to dry-cured meat products), (2) dendritic salt crystals riddled with pores, and (3) Maldon salt—hollow, pyramid-shaped flakes from Essex, England (Kilcast & den Ridder, 2007). The high surface/volume ratios of all these shapes enhance the salt dissolution rate, enabling water molecules in saliva to break the ionic bonds of NaCl crystals more readily.

Sodium in pizza crusts can also be reduced by using larger crystals and less overall salt. Large salt crystals have comparatively small surface/volume ratios. Mueller et al. (2016) found that if coarse-grained salt (0.4–1.4 mm) was added to pizza dough 30s before the end of mixing time, the salt crystals underwent little dissolution. They remained as salty spots in the crust, readily discernable during consumption. Taste tests showed that a 25% total salt reduction could be achieved without affecting the perception of saltiness.

A high-tech method of reducing sodium in food was recently developed by the Japanese pharmaceutical company Kirin. It created computer-controlled chopsticks that use a tiny electrical current to draw extra sodium ions from food and deposit them onto the eater's tastebuds. Studies suggest that total sodium can be reduced by a third without affecting apparent saltiness. Inspired by these chopsticks and by a concern for those who should limit their sugar intake, researchers recently created a spoon (dubbed Sugarware) with bumps on its bottom that stimulate taste buds to produce a sensation of sweetness.

7.5 Sugar and Candy Making

There are several varieties of candy made from sugary solutions. The initial step in preparing most candies is to bring a sugary aqueous solution to boiling. Hot water can dissolve more sugar than cold water; greater thermal energy aids agitated water molecules in severing the hydrogen bonds that hold separate sucrose molecules together. Cooling of the sugary solution produces a viscous syrup.

Rock candy has coarse sucrose crystals, formed by allowing the sugar syrup to cool slowly over several days. In many cases, a string (or stick) is inserted into the syrup to act as a nucleation site (Fig. 7.9). As the syrup cools, the solution becomes supersaturated, and sucrose begins to nucleate on the string; relatively few nuclei form away from the string. With additional slow cooling, freely floating sucrose molecules in the syrup attach themselves to existing nuclei, forming a relatively small number of large crystals attached to the string.

Fudge contains small sugar crystals, produced by stirring the syrup while it cools quickly, no strings attached. Stirring increases the kinetic energy of the dissolved sucrose molecules; they bump into each other and form numerous small nuclei throughout the solution. The nuclei are generally of similar size and shape. Their high surface/volume ratios facilitate nucleation. The polar sucrose molecules attach themselves to the growing nuclei; the negatively charged areas near the oxygen atoms of one molecule attract the positively charged areas near the hydrogen atoms of another. The absence of a string inside the syrup and the rapid cooling rate serve to inhibit the formation of large crystals.

Fig. 7.9 Rock candy on a string. *Image* from Miansari66

Fig. 7.10 Cotton candy extruded from a cotton candy machine and spun around a stick at a fair. *Image* from FocalPoint

A third type of sugar candy is not crystalline at all; examples include <u>glass candy</u> and <u>cotton candy</u>. Glass is an amorphous substance with no long-range crystalline order. If a sugar syrup is quenched, it will solidify into glass. Sugar glass is sometimes used for decorating cakes, cupcakes, and doughnuts; recipes can be found on the internet. Sugar glass has also been used in movies and on stage to simulate shattering window glass during fight scenes. Cotton candy (candy floss in the U.K. and Ireland; fairy floss in Australia) is made from melted granulated sugar sprayed through the narrow nozzles of a cotton candy machine into room-temperature air. The fine spray has a very high surface/volume ratio and solidifies quickly into thin filaments. These fine strands of sugar glass are spun rapidly around a stick (Fig. 7.10). The final product resembles cotton bolls. Food coloring and flavoring are often added. Cotton candy, wrapped around sticks or paper cones or stuffed in small plastic bags, is a popular confection at fairs and carnivals, but like all sugar candy, is frowned upon by the World Health Organization and killjoy dentists. WHO recommends limiting free sugars to less than 5% of total energy intake to diminish the risk of dental cavities.

Natural silicate analogs of cotton candy are strands of Pele's hair (Fig. 7.11), brown fibers of basaltic glass formed from quenched droplets of sprayed lava, spun or stretched into filaments. Strands range in thickness from 1 to 300 μm (Duffield et al., 1977). Wind currents can transport strands of Pele's hair more than 10 km from the site of a volcanic eruption. The glassy filaments are named after Pele, the Hawaiian goddess of volcanoes.

7.6 Evaporation

Evaporation is the transfer of molecules from the surface of a liquid to a surrounding gas. Many factors influence the evaporation rate:

Fig. 7.11 Pele's hair, inedible volcanic cotton candy. These thin strands of basaltic glass form from quenched droplets of sprayed lava. The round glass lens is approximately 2 cm across. *Image* by Cm3826

1. Vapor pressure. If the vapor pressure exerted by the gas on the liquid is high, more energy is required for molecules to escape the liquid's surface and enter the gas phase.
2. Temperature of the liquid. At higher temperatures, molecules in a liquid have more kinetic energy; they collide with one another vigorously, transferring energy from one molecule to the next. When a molecule situated near the liquid surface receives enough kinetic energy to overcome the vapor pressure, it escapes the liquid and enters the gas. This removes energy from the liquid, causing it to cool (a process known as evaporative cooling). The amount of heat that must be added to a liquid to transform a given amount of the liquid into a gas is known as the heat of vaporization.
3. Concentration of the evaporating substance in the gas. If there are only small amounts of the evaporating substance in the gas, evaporation will be faster. In contrast, if the concentration level of the evaporating substance in the gas is near saturation (e.g., in a closed system), evaporation from the liquid will be sluggish. (This can be viewed as an effect of differences in chemical potential—substances tend to move from regions of high concentration to those of low concentration until equilibrium is reached.) If the gas has a high flow rate above the liquid, the liquid will evaporate faster; this scenario is essentially the same as replacing a stagnant gas with one in which the concentration of the evaporating substance is lower.
4. Strength of intermolecular forces in the liquid. If the chemical bonds holding molecules in the liquid together are strong, more energy is required for the molecules to overcome the vapor pressure and enter the surrounding gas.
5. Surface area of the liquid. The greater the surface area of the liquid exposed to the gas, the higher the rate of evaporation. There is a one-to-one correlation: A 50% reduction of the surface area of a liquid in a reservoir results in a 50% reduction in the volume of liquid evaporated.

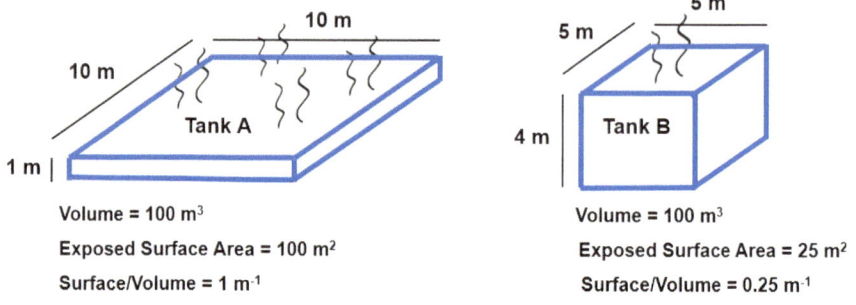

Fig. 7.12 Two open-air water tanks of identical volume but with different dimensions. The surface/volume ratio of Tank B is four times lower than that of Tank A and will allow only 25% as much evaporation from the upper surface. After McJannet et al. (2008)

A simple example (Fig. 7.12): There are two open-air containers holding an equal volume of water ($100 \, m^3$). Tank A is wide and shallow: 10 m wide, 10 m long, 1 m deep. It has an upper surface area of $100 \, m^2$ and a surface/volume ratio of $1 \, m^{-1}$. Tank B is short and deep: 5 m wide, 5 m long, 4 m deep. It has an upper surface area of $25 \, m^2$ and a surface/volume ratio of $0.25 \, m^{-1}$. Under identical atmospheric conditions, the amount of water evaporated from Tank B would be four times lower than from Tank A.

If drought conditions are present in an urban area, the surface/volume ratio of such open-air water-storage containers would be an important consideration in water-conservation efforts. The government of Queensland, Australia partnered with local universities to produce a report addressing this very problem (McJannet et al., 2008). Unfortunately, the authors were unable to find viable large-scale solutions to the problem of evaporation and suggested it was best to use small storage containers.

7.7 Osmosis

Living cells are completely bounded by a membrane—a two-layered flat sheet that serves as the interface between the interior of the cell and the external aqueous environment (Fig. 7.13). The membrane is made of large lipid molecules along with various proteins and cholesterol; it is insoluble in water and semipermeable. It allows the passage into and out of the cell of particular ions or molecules (solutes) as a function of temperature, pressure, electric charge, solute size, and the concentration levels of the different solutes within and outside the cell. Membrane proteins facilitate the transport of solutes across the boundary. The membrane typically allows the transport of small molecules (e.g., oxygen, carbon dioxide, molecular nitrogen) into the cell, but bars entry to many macromolecules.

Unlike the cells of animals, plant cells have a cell wall (Fig. 5.31), typically rich in cellulose and located just outside the membrane. Cell walls provide structural

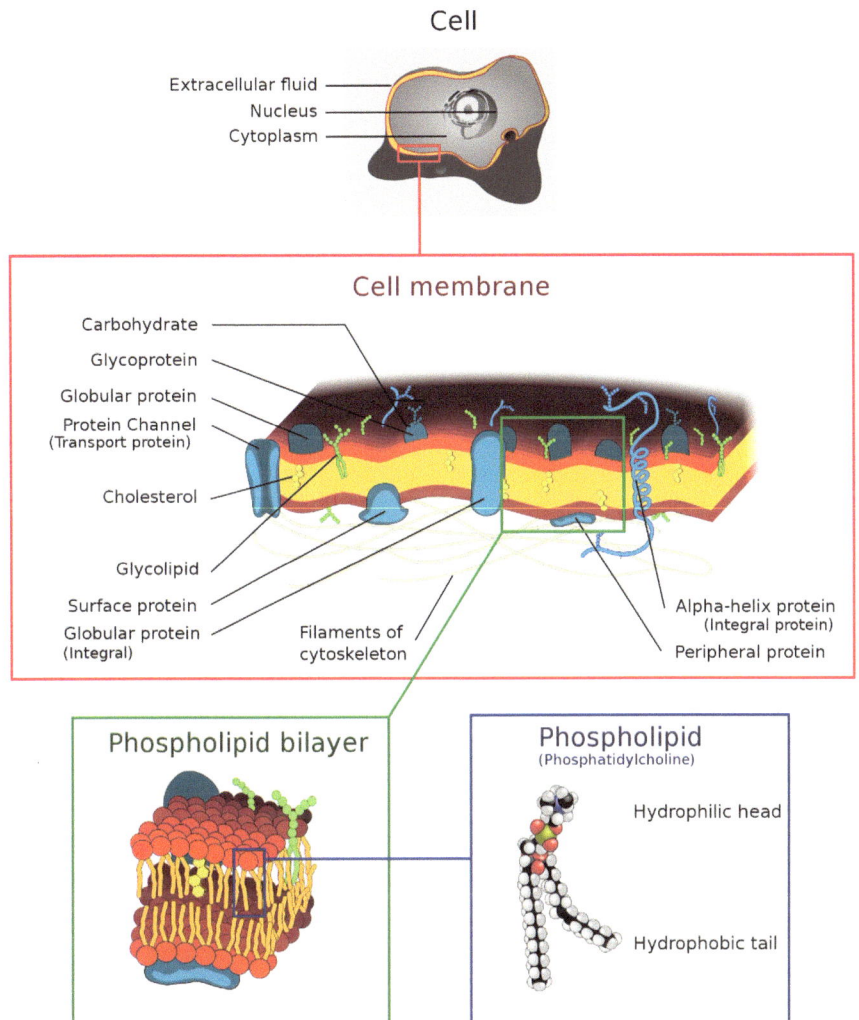

Fig. 7.13 The cell membrane. *Image* from Wikimedia Common

support for the cell; the walls become rigid after water diffuses through the cell membrane. This rigidity (or turgidity) prevents the cell from becoming too distended and bursting. Turgor is the pressure exerted on the cell wall by the fluid inside the cell; it is the property that makes plant tissues rigid.

Osmosis is the process of diffusion of solutes across the cell membrane from regions of high concentration to those of low concentration. The solutes are transported in the same direction that would be required to achieve equal concentrations on both sides of the boundary.

Fig. 7.14 Mean percentage increase in mass over the first hour for potato cubes immersed in distilled water. The large cube (with its small surface/volume ratio) absorbs water at a much slower rate. After Barrett (1984)

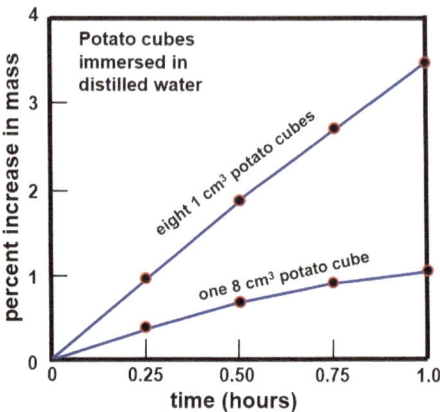

A class-room experiment reported in the *Journal of Biological Education* serves to illustrate the role played by the surface/volume ratio in plants in regulating osmosis (Barrett, 1984). Students fashion peeled potatoes into (1) eight small cubes (each with dimensions of 1 cm × 1 cm × 1 cm) and (2) one large cube (2 cm × 2 cm × 2 cm). The small cubes have a combined surface area of 48 cm^2, a combined volume of 8 cm^3, and a surface/volume ratio of 6 cm^{-1}. The large cube has a surface area of 24 cm^2, a volume of 8 cm^3, and a surface/volume ratio of 3 cm^{-1}. Because potatoes are of uniform density, over many experiments, the mean collective weight of the eight 1-cm cubes should be about the same as the one 2-cm cube.

All potato cubes were weighed before the experiment, then immersed in separate beakers of distilled water. Every 15 min, over the course of one hour, the cubes were patted dry and reweighed. The results are plotted in Fig. 7.14.

It is clear the set of small potato cubes absorbs water at a much faster rate than the single large cube. Over the hour, the difference in absorption progressively increases from a factor of 2.6–3.3. This is a function of the difference in the surface/volume ratio: the ratio for the small cubes is twice that of the large cube (6 cm^{-1} vs. 3 cm^{-1}).

Water is entering potato cells via osmosis through the cell membrane, causing the cells to become fully turgid. Because the small cubes have twice as many cells in direct contact with the water at the surface of the potato, the small cubes absorb more water and at a faster absorption rate. Interior potato cells may fail to absorb as much water because water pathways are blocked by the turgor of the cells nearer the surface. Because a much higher proportion of cells are in the interior of the large cube (due to its lower surface/volume ratio), this cube absorbs water much more slowly. Over the duration of the experiment, the lone large cube absorbs far less total water than the set of small cubes.

7.8 Grain Elevators

Since the adoption of agriculture 10,000–12,000 years ago, humans have been storing grain. There were unintended consequences beyond those involving social reorganization. The grain attracted mice and the mice attracted cats. Humans were pleased with feline pest control and let the cats hang around. This tolerance led eventually to the wide-spread domestication of cats. In Egypt, it led to reverence. In 1888 archaeologists found a cat cemetery near the small village of Beni-Hassan with 80,000 mummies of cats and kittens. Paintings of cats decorated the walls.

The necessity of storing grain found its way into the Bible. In Genesis 41:34–36 (NIV), Joseph tells Pharoah: "Let Pharaoh appoint commissioners over the land to take a fifth of the harvest of Egypt during the seven years of abundance. They should collect all the food of these good years that are coming and store up the grain under the authority of Pharaoh, to be kept in the cities for food. This food should be held in reserve for the country, to be used during the seven years of famine that will come upon Egypt, so that the country may not be ruined by the famine."

The modern solution for storing large quantities of grain is the grain elevator (Fig. 7.15). These structures are typically 20–40 m tall and located near farms; most are in the vicinity of transportation routes (highways, rail lines, navigable rivers). Farmers load their sacks of grain onto trucks and drive to the local grain elevator. The trucks are weighed, the grain offloaded and inspected. The moisture content of the grain should be about 15%. If the grain is too wet, mold could form; if the grain is too dry, it could age prematurely. Once the grain passes inspection, it is emptied into a pit beneath the slotted work floor of the elevator. Buckets attached to a continuous belt scoop the grain up and haul it to the top of the elevator where it is deposited into silos. The trucks are weighed again to determine the amount of offloaded grain.

But there is a problem with grain elevators. They can explode. On average, there are 10.6 agricultural dust explosions in the United States every year, causing 1.6 deaths and 12.6 injuries; millions of dollars of equipment are destroyed (Jones, 2017). An explosion at the Husted Mill and grain elevator in Buffalo, New York in 1913 caused 33 deaths and 80 injuries; one at the Westwego grain elevator in Louisiana in 1977 caused 36 deaths and 13 injuries; that same year, an explosion at the Galveston grain elevator in Texas killed 20 people.

There are four principal conditions that must be present for a grain elevator to explode: (1) fuel (combustible grain dust particles of sufficiently high concentration), (2) oxygen (ambient air), (3) confined space (e.g., the grain elevator, silo, drag conveyor system, basement tunnel), and (4) a source of ignition (e.g., lightning, static electricity, spark from electrical equipment or a welding torch, overheated equipment).

Dust explosions tend to occur at sites where the grain is being moved, e.g., bucket elevators and conveyors. Heat can accumulate gradually via friction (as in a slipping conveyor belt) or can be produced rapidly by an impact (e.g., from a dislodged piece of broken equipment) or an electrical spark. During transportation, small dust

Fig. 7.15 Grain elevator with silos on a farm in Israel. *Image* from David Shankbone

particles (2–20 μm across) could break off larger grains and become suspended in the confined airspace, possibly in the vicinity of a heat source.

A primary explosion in the silo or elevator often leads to secondary explosions in adjacent areas. Shock waves from the primary explosion deposit dust in nearby areas that are ignited by the flame following in the wake of the shock waves. Chain reactions can occur in which secondary explosions produce additional explosions. Primary explosions have been reported to generate pressures of $1400\,kg/m^2$; pressures from secondary explosions can exceed $56{,}000\,kg/m^2$ (Jones, 2017).

Agricultural materials that have produced dust explosions include various grains, fluor, starch, sugar, and coffee. These products are potentially dangerous (even if they are not ingested). The amount of heat that can be produced by an explosion (heat of combustion) is 18% greater for sucrose and starch than for coal (470 kJ vs. 400 kJ per mole of molecular oxygen consumed) (Eckhoff, 2003).

The combustion reaction for a hydrocarbon is:

$$C_xH_x + zO_2 \rightarrow xCO_2 + (y/2)H_2O + heat$$

where $z = x + (y/4)$. Hydrocarbon fuels include members of the alkane series (general formula C_nH_{2n+2}) in which the C and H atoms are arranged in a tree structure: methane (CH_4), ethane (C_2H_6), propane (C_3H_8), butane (C_4H_{10}), pentane (C_5H_{12}), hexane (C_6H_{14}), heptane (C_7H_{16}), and octane (C_8H_{18}).

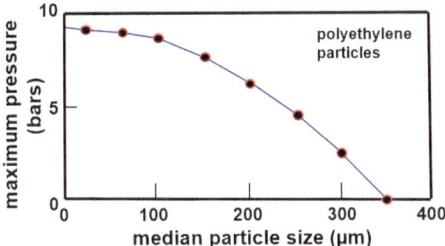

Fig. 7.16 Maximum pressure produced during closed-bomb explosion experiments for polyethylene $(C_2H_4)_n$ particles of different sizes. The smallest particles (with the highest surface/volume ratios) produce greater maximum pressures. After Eckhoff (2003)

In the case of the organic compound cellulose, $(C_6H_{10}O_5)_n$, [not a hydrocarbon because it contains oxygen], the simplified combustion reaction is:

$$C_6H_{10}O_5 + 6O_2 \rightarrow 6CO_2 + 5H_2O + heat$$

The reactivity of dust is a function of particle size. Smaller grains, with their higher surface/volume ratios, are more reactive than larger grains. In comparable concentrations, smaller grains produce higher maximum explosive pressures (Fig. 7.16), whether the grains are organic materials or metal powders.

The minimum explosive concentration of dust particles ranges from about 50 to 150 g per cubic meter. Smaller particles require lower concentrations for an explosion to occur. That is because the combined surface area of small particles available for combustion far exceeds that of large particles of the same total mass. For example, a 100-g sphere of a combustible substance with a density of 1 g cm^{-3} would be 5.76 cm in diameter and have a surface area of 104 cm^2. If this sphere were broken into 50-μm spherical, fluor-size, particles (with no loss of mass), the collective surface area would increase by a factor of 1150 to about 120,000 cm^2.

For any combustible material, smaller particles will ignite at a lower temperature than larger particles, consistent with the greater ability of molecular oxygen to diffuse into small particles (again, due to their higher surface/volume ratios). However, below a certain particle size, further decreases do not lower the minimum ignition temperature (MIT). Data from Zhao et al. (2020) show that the MIT decreases steadily for progressively smaller dust particles (with their steadily increasing surface/volume ratios) that are suspended in a cloud until small sizes are reached. After this point, the MIT remains constant (Table 7.1).

Smaller particles with their higher surface/volume ratios also produce greater rates of increase in maximum pressure during explosions (Fig. 7.17). This applies to organic materials as well as metal powders.

Early grain elevators were constructed from framed or cribbed wood; many were destroyed by fire. Modern grain elevators and silos are made from steel or reinforced

Table 7.1 Minimum ignition temperature (MIT) for particles of different sizes suspended in a dust cloud

Mesh size	Particle size range (μm)	Surface/volume[a]	MIT (°C)
80	180–1250	0.0333	490
100	154–180	0.0390	470
120	120–154	0.0500	450
140	109–120	0.0550	440
160	96–109	0.0625	430
180	80–96	0.0750	430

Data from Zhao et al. (2020). Particles were sorted by standard sieve into the listed mesh sizes. The listed minimum size of the range is the smallest axis of a potentially irregularly shaped particle
[a]The surface/volume ratios are calculated by assuming the particles are spheres with a radius equivalent to half the minimum-size particle that makes it through the mesh.

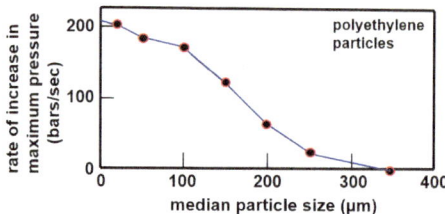

Fig. 7.17 Rate of increase in maximum pressure (bars/second) produced during closed-bomb explosion experiments for polyethylene $(C_2H_4)_n$ particles of different sizes. The smallest particles (with the highest surface/volume ratios) produce higher rates of increase in maximum pressure. After Eckhoff (2003)

concrete. They are less likely to burn down but are still susceptible to internal explosions. More than 250,000 people in the United States work at grain elevators and mills. Insurance costs are high.

References

Barrett, D. R. B. (1984). Osmosis and surface area to volume ratio. *Journal of Biological Education, 18*, 273–274.

Blanc, P.-L. (2002). The opening of the Plio-Quaternary Gibraltar Strait: assessing the size of a cataclysm. *Geodinamica Acta, 15*, 303–317.

Duffield, W. A., Gibson, E. K., & Heiken, G. H. (1977). Some characteristics of Pele's hair. *Journal of Research U. S. Geological Survey, 5*, 93–101.

Eckhoff, R. K. (2003). *Dust Explosions in the Process Industries* (3rd ed.). Gulf Professional Publishing, 729 pp.

He, F. J., & MacGregor, G. A. (2008). A comprehensive review on salt and health and current experience of worldwide salt reduction programmes. *Journal of Human Hypertension, 23*, 363–384.

Jones, C. (2017). *Preventing grain dust explosions.* Division of Agricultural Sciences and Natural Resources, Oklahoma State University. Information on: http://osufacts.okstate.edu

Kilcast, D., & den Ridder, C. (2007). Sensory issues in reducing salt in food products. In D. Kilcast & F. Angus (Eds.), *Reducing Salt in Foods: Practical Strategies* (pp. 201–220). Woodhead Publishing.

McJannet, D., Cook, F., & Burn, S. (2008). Evaporation reduction by manipulation of surface area to volume ratios: Overview, analysis and effectiveness. *Urban Water Security Research Alliance Technical Report, 8,* 21 pp.

Mueller, E., Koehler, P., & Scherf, A. (2016). Applicability of salt reduction strategies in pizza crust. *Food Chemistry, 192,* 1116–1123.

Zhao, J., Tang, G., Wang, Y., & Han, Y. (2020). Explosive property and combustion kinetics of grain dust with different particle sizes. *Heliyon, 6.* https://doi.org/10.1016/j.heliyon.2020.e03457

Chapter 8
Ecology

8.1 Wildfires

Wildfires are uncontrolled fires that consume combustible vegetation and spread across the landscape (Fig. 8.1). The <u>wildfire front</u> is the advancing boundary between flaming vegetation and unburned material. Near the front, unburned vegetative matter is heated and dried by thermal radiation from nearby flames and by heat transfer from superheated air. The front can race through forests at ~10 km per hour and over grasslands at ~20 km per hour.

These fires can be devastating. In the summer of 2019–2020, wildfires in Australia burnt 60,000 to 100,000 km^2 (including 21% of temperate forest cover), killed or displaced nearly three billion non-insectoid animals (2.46 billion reptiles, 143 million mammals, 180 million birds, 51 million frogs), produced huge amounts of carbon dioxide, and (aided by solar heating) propelled enormously wide smoke plumes to record heights (more than 31 km) into the mid stratosphere. Ecological damage includes destroyed and degraded habitats, and the endangerment of several hundred native vertebrate and invertebrate species.

<u>Fuel loading</u> is the amount of fuel available in a region. In the United States, this parameter is usually measured in tons per acre. Grassland has a lower loading value than downed trees.

Fires can be categorized by the type of fuel consumed and the fuel location:

1. *Ground fires* are fueled by peat, decomposing roots, and decaying organic matter. They spread horizontally and may smolder and burn for several weeks.
2. *Surface fires* (a.k.a. *crawling fires*) are fueled by vegetation at the forest floor including leaves, grass, shrubs, needles, twigs, branches, timber litter, moss, and lichens. They also spread horizontally.
3. *Ladder fires* are fueled by vegetative matter at intermediate heights (the understory) including small trees, tall shrubs, low branches, vines, and climbing plants. They spread vertically.
4. *Canopy fires* are fueled by high-level vegetation at the canopy and emergent layers, including tall trees, high-level branches, and crowns.

© The Author(s), under exclusive license to Springer Nature Switzerland AG 2023
A. E. Rubin, *Surface/Volume*, https://doi.org/10.1007/978-3-031-23749-2_8

Fig. 8.1 The Rim Fire started in Stanislaus National Forest in California on 17 August 2013 and burned 1041 km^2 of forest near Yosemite National Park. At the time, it was the second-largest wildfire in California history. *Image* from U.S. Department of Agriculture

Fuel availability is a measure of how readily an available fuel will burn. It is a function of moisture content as well as fuel size, fuel shape and fuel surface/volume ratio. Fuels are categorized by how long it would take them to dry out sufficiently to burn readily: 1-h and 10-h fuels are those mainly responsible for ignition and initial fire spread. Dead needles, mulch, leaves, grass, and litter have very high surface/volume ratios, dry out quickly, and ignite readily; they are typical of 1-h fuels (defined as those less than 0.25 in. (0.635 cm) in diameter). Ten-hour fuels are those with diameters between 0.25 and 1.0 in. (0.635–2.54 cm); they include roundwood and thick litter. One-hundred-hour fuels are 1–3 inches in diameter (2.54–7.62 cm); 1000-h fuels are 3–8 in. in diameter (7.62–20.32 cm). The physically larger individual fuel sources are harder to ignite because of their low surface/volume ratios, but once set aflame, they will burn for longer periods (due to their higher volume and greater mass).

A standard fuel log for an American fireplace is a right circular cylinder 16 inches long and 6 inches thick (40.64 × 15.24 cm). The surface area of an ideal log is 2311 cm^2, its volume is 7413 cm^3, and its surface/volume ratio is 0.312 cm^{-1}. If this log is split into 16 identical cylindrical disks, each 1 inch long and 6 inches thick (2.54 × 15.24 cm), with no loss of mass, then the combined surface area would be 7783 cm^2. The volume would be unchanged, and the new surface/volume ratio would be 1.05 cm^{-1} (a factor of 3.37 higher). The divided logs would have a much higher combined surface area and a much higher collective surface/volume ratio, thus greatly increasing fuel availability (Fig. 8.2). This is not just a mathematical exercise. Whenever a burning log splits in the fireplace, the flames soon grow higher; more heat is generated per minute as the wood is consumed at a more rapid rate.

Fig. 8.2 A split log has more surface area than an intact log. The increased fuel availability results in more flames and a more rapid burning rate

It is important to understand that wildfires need not be unmitigated disasters. Normal, low-intensity, wildfires can be beneficial to ecosystems that developed in regions where such fires are common. Complex early seral forests (an intermediate growth stage prior to the re-formation of a closed forest canopy) develop after a fire has destroyed an old-growth standing forest. These regions have a rich biodiversity due to the heterogeneity introduced by remaining trees, charred logs, fungi, sprouting seedlings, and the animals occupying these ecological niches.

Some plants called pyrophytes (fire lovers) depend on the occasional fire to survive. After a fire, young longleaf pine trees develop a rough bark, resistant to brown spot fungus; mountain ash needs fire to prepare seedbeds for its progeny; and firewood seed does not sprout until the litter and soil around it are removed by flame (Pyne, 2009). Low-intensity seasonal wildfires leave some trees alive and decrease the fuel load by eliminating 1- and 10-h fuels (the small-sized vegetative matter with high surface/volume ratios). The remaining fuels are of larger size with small surface/volume ratios; they are more difficult to ignite.

The United States federal government established an aggressive and misguided wildfire policy in the period 1905–1911. It was maintained for many decades. Wildfires were to be suppressed to protect precious natural resources and ever-expanding human communities. But there was no ancillary program to reduce the vegetative fuel that accumulated gradually once low-intensity seasonal wildfires were systematically extinguished (Busenberg, 2004). The result was a massive increase in fuel load. One ponderosa pine forest site in Arizona experienced an increase in tree density by a factor of eight between 1883 and 1995 (Covington & Moore, 1994).

When fires break out after decades of fire suppression, the immense fuel loads produce high-intensity fires; they kill nearly every tree, destroy the understory, and significantly disturb the soil. The situation is worsened by the effects of climate change—the fuel load is dried further and becomes easier to ignite.

Record-setting fires of great magnitude have become commonplace in the American West. In 2020, California had nearly 10,000 fires that burned through 4.4 million acres (roughly 18,000 square kilometers). A single fire—The August Complex Fire, informally dubbed a "gigafire"—burned more than one million acres (~4000 km²), destroyed 10,000 structures, and racked up 12 billion dollars in damages (Fig. 8.3).

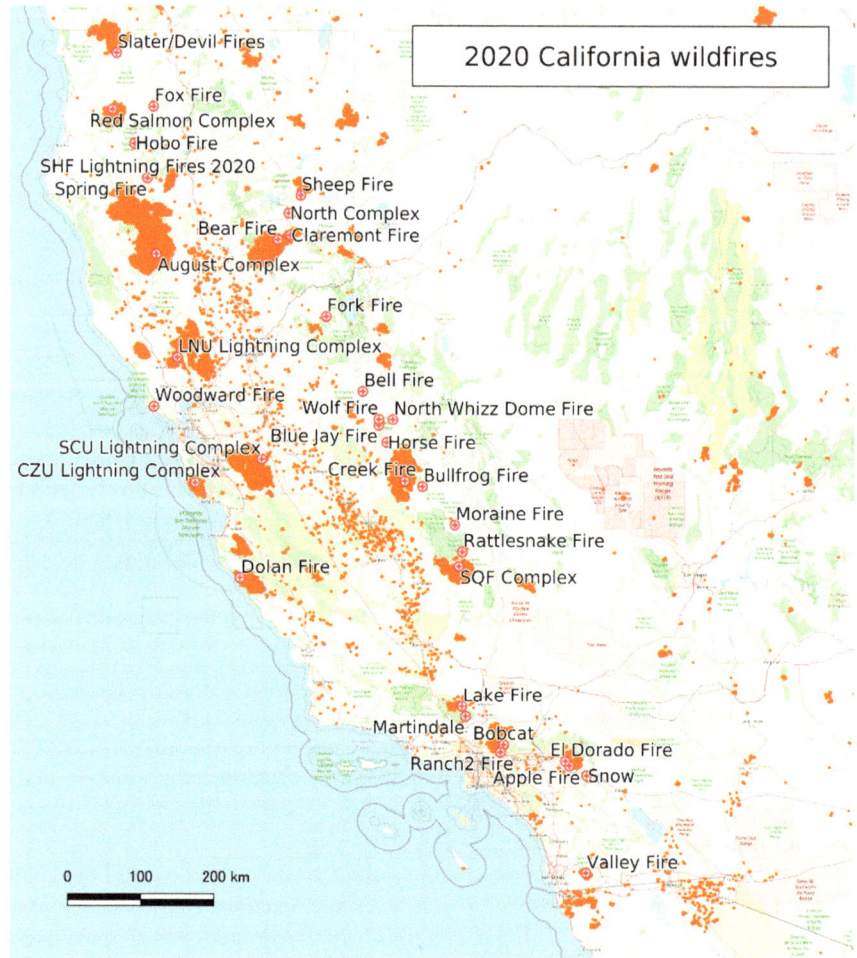

Fig. 8.3 Wildfires in California in 2020. *Data* from the U.S. Forest Service. *Diagram* from Phoenix7777

Gross mismanagement, coupled with climate change, have caused significant threats to public safety. There have been loss of life, massive costs of firefighting, extensive property damage, appreciable diminishment of commodity values (e.g., timber, tourism), severely compromised ecosystems, imperiled watersheds, reduced air quality, and large costs in rehabilitating forests and human communities.

8.2 Coral Reefs

Coral reefs are structures built by colonies of coral and cemented by calcium carbonate ($CaCO_3$) secreted by the coral. Although reefs constitute less than 0.1% of the total ocean area, they harbor about 25% of marine species: fish, sponges, marine worms, crustaceans (e.g., crab, lobsters, shrimp), echinoderms (e.g., star fish, sea urchins, sand dollars), mollusks (e.g., squid, octopus, cuttlefish), tunicates (e.g., sea squirts), and cnidarians (e.g., sea anemones, coral, jellyfish). These reef inhabitants are an important local food source; globally, reefs provide a livelihood for 125 million people (Lesser, 2011).

Corals manifest bright colors (Fig. 8.4); they live in a symbiotic relationship with dinoflagellates (zooxanthellae)—microscopic photosynthetic algae responsible for coral pigmentation. The algae supply the products of photosynthesis (glucose, glycerol, amino acids) to the coral; the coral uses these compounds to make proteins, fats, carbohydrates, and calcium carbonate. The coral serves as a protective harbor for the algae. Because of their reliance on photosynthetic algae, corals generally grow in warm, clear, shallow waters of low turbidity. Such conditions optimize the amount of sunlight that reaches the algae.

However, if water temperatures rise by even one or two degrees Celsius, the zooxanthellae begin to produce reactive oxygen compounds (e.g., superoxide radical—O_2^-; hydrogen peroxide—H_2O_2; hydroxyl radical—OH^-+; hydroxyl ion—OH^-) that can damage coral lipids, proteins, and DNA (Lesser, 2011). The coral reacts

Fig. 8.4 Healthy coral at Lodestone Reef in Australia. *Image* by Holobionics

Fig. 8.5 Bleached branching coral at the Great Barrier Reef in Australia. *Image* by J. Roff

by expelling the algae from their tissues. Pigmentation is lost; the corals turn pale. This phenomenon, known as coral bleaching, is a sign of diminished coral health (Fig. 8.5). Unless water temperatures decline, the coral will bar the algae from reentry and the coral will die. Other potential causes of bleaching include a rise in ocean acidity (due to increased amounts of dissolved carbon dioxide), changes in salinity and sedimentation, and increased exposure to ultraviolet radiation. Since the 1950s, global coral reef cover has dropped by 50% (Eddy et al., 2021).

Not all coral genera are equally susceptible to bleaching. Fine-structured, branching, and flat, leaf-like corals (e.g., *Acropora, Millepora, Montipora*) bleach easily. Massive corals with thick tissues (e.g., *Cyphastrea, Diploastrea, Favia, Goniastrea*) can acclimatize to elevated temperatures more readily; they undergo less bleaching and are better able to survive bleaching episodes. Possible reasons for the hardiness of these massive corals include superior protein repair processes and more thermally tolerant varieties of embedded zooxanthellae.

There are also differences between these branching and massive corals in surface/volume ratio. Fine-structured corals have high surface/volume ratios. It seems plausible that the zooxanthellae-produced toxins rapidly penetrate the entire coral body, causing widespread tissue damage, leading to algae expulsion and coral bleaching. In addition, during periods of high solar irradiance, these fine-structured corals receive more light per unit volume. Most photons are absorbed by the pigmented corals; unabsorbed photons are scattered throughout the structure, enhancing the opportunity for more of it to reach the zooxanthellae (Enriquéz et al., 2005). When corals are stressed (e.g., during periods of elevated water temperatures) and the pigmentation is reduced, the photosynthetic capacity of the algae decline. As concluded by Enriquéz et al. (2005), the photoprotective mechanisms corals use to dissipate excess thermal energy "may be overwhelmed.... resulting in coral bleaching and mortality."

In contrast, massive corals have low surface/volume ratios. They receive less light per unit volume, diminishing the chances their photoprotective mechanisms will be overwhelmed during times of elevated water temperature. It is also possible that toxins produced by stressed zooxanthellae in these corals are impeded from penetrating the entire coral mass. This could delay or prevent algae expulsion and coral bleaching.

References

Busenberg, G. (2004). Wildfire management in the United States: The evolution of a policy failure. *Review of Policy Research, 21*, 145–156.

Covington, W. W., & Moore, M. M. (1994). Postsettlement changes in natural fire regimes and forest structure: Ecological restoration of old-growth ponderosa pine forests. *Journal of Sustainable Forestry, 2*, 153–181.

Eddy, T. D., Lam, V. W. Y., Reygondeau, G., Cisneros-Montemayor, A. M., Greer, K., Palomares, M. L. D., Bruno, J. F., Ota, Y., & Cheung, W. W. L. (2021). Global decline in capacity of coral reefs to provide ecosystem services. *One Earth, 4*, 1278–1285.

Enriquéz, S., Méndez, E. R., & Iglesias-Prieto, R. (2005). Multiple scattering on coral skeletons enhances light absorption by symbiotic algae. *Limnology and Oceanography, 50*, 1025–1032.

Lesser, M. P. (2011). Coral bleaching: Causes and mechanisms. In Z. Dubinsky & N. Stambler (Eds.), *Coral reefs: An Ecosystem in Transition* (pp. 405–419). Springer.

Pyne S. J. (2009) How plants use fire (and are used by It). NOVA Online. https://web.archive.org/web/20090808123751/http:/www.pbs.org/wgbh/nova/fire/plants.html

Chapter 9
Manufacturing

9.1 Artificial Bones

There are two types of *natural* bone compact bone (a.k.a. cortical bone), accounting for 80% of the bone mass in humans, and spongy bone (a.k.a. cancellous bone or trabecular bone), accounting for 20% (Fig. 9.1).

As its name implies, compact bone is dense (~2 g cm^{-3}), consisting of hard bone with ~30 vol.% vascular channels. It is the strong bone component, providing structure and support for the body; it facilitates movement via connective tissues (e.g., tendons and ligaments). Compact bone is composed of osteons (quasi-cylindrical units 0.25–0.35 mm in diameter) that are made of different cell types including osteocytes (that maintain bone mass) and osteoblasts (that form new bone). Osteons consist of concentric shells of compact bone tissue encircling a central canal (a.k.a. the haversian canal). The canals allow nutrients from the body to enter the bone and wastes to leave. Compact bone is much less porous than spongy bone and thus has a lower surface/volume ratio. Blood vessels in the compact bone deliver nutrients (glucose, amino acids, lipids, and trace elements) to the spongy bone.

Spongy bone is more porous, much less dense (0.47 g cm^{-3}; Li & Aspden, 1997), and more flexible than compact bone; it consists of 20 vol.% bone and 80 vol.% red bone marrow and fat. It contains less calcium and more water than compact bone. There are no osteons. It is composed of a lattice network of tissue (creating a spongy appearance) and is responsible for transferring mechanical loads to compact bone. Spongy bone contains numerous osteocytes that, in humans, produce about two million red blood cells per second.

Bones are made mainly of hydroxylapatite ($Ca_5(PO_4)_3(OH)$) and collagen fibers (elongated structural proteins bound in a triple helix). Although bones are strong, they are susceptible to fracturing. Small fractures can be auto-repaired by bone resorption (via osteoclast cells) and bone synthesis and mineralization (via osteoblast cells). But if a person suffers a major bone injury (e.g., compound fractures from a car accident) or if the bone has been appreciably degraded by disease (e.g., osteoporosis, osteoarthritis, bone cancer), a surgeon may remove affected bone and replace it

A. E. Rubin, *Surface/Volume*, https://doi.org/10.1007/978-3-031-23749-2_9

Compact Bone & Spongy (Cancellous Bone)

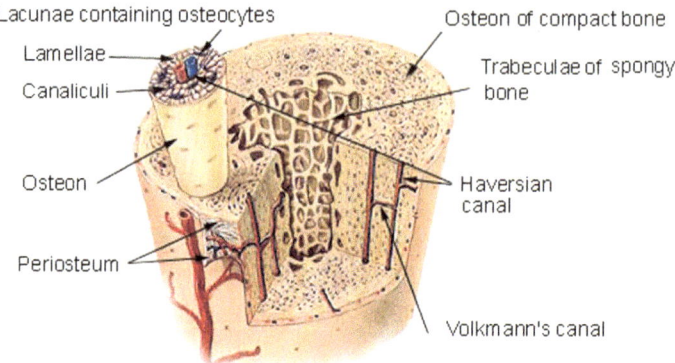

Fig. 9.1 Illustration of compact bone and spongy bone. *Image* from the U.S. National Cancer Institute's Surveillance, Epidemiology and End Results (SEER) Program

with autografts (bone from the same individual), allografts (bone from a donor), or artificial bone (synthetic grafts). There is plenty of work for orthopedic surgeons—more than two million bone grafts are performed worldwide each year.

Artificial bones made from a composite of hydroxylapatite and chitosan (long-chain carbohydrates derived from shellfish) have numerous advantages. These include biocompatibility, high osteoconductivity (the ability to facilitate bone growth on the artificial bone surface), thermodynamic stability, mechanical strength, high porosity, straightforward surgical implantation, and the avoidance of additional surgery involving bone excision and collection for bone grafts (e.g., Hoppe et al., 2011; Saijo et al., 2016). The artificial bones are molded into porous scaffolds that, after implantation, allow vital molecular oxygen (O_2) and nutrients to reach the living cells that grow on the scaffold structure.

Nguyen et al. (2019) tested two common scaffold types to determine which one allowed more opportunities for collisions between O_2 and bone cells. They examined (1) fiber meshes constructed from polymer microfibers and (2) porous foam scaffolds (Fig. 9.2), measuring how far molecular oxygen could travel before it was totally consumed by the cells. The higher the surface/volume ratio of the scaffold, the faster O_2 was absorbed. This is because a higher exposed surface area provides a greater chance for O_2 molecules to collide with scaffold material. Fiber scaffolds have a higher surface/volume ratio than foam scaffolds, leading to more rapid consumption of O_2 by bone cells. Thus, fiber scaffolds seem to be the better choice for artificial bone implantation.

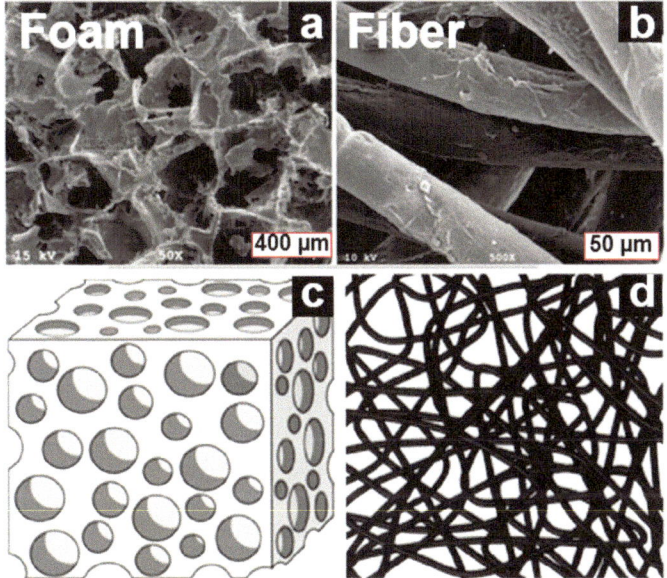

Fig. 9.2 Two common types of artificial bone scaffolds: **a** Scanning electron microscope (SEM) image of a porous foam scaffold. **b** SEM image of a fiber mesh scaffold. **c** Illustration of large-scale structure of foam scaffold with spherical pores. **d** Illustration of large-scale structure of fiber scaffold with mesh-shaped pores. All images modified from Nguyen et al. (2019)

9.2 Artificial Lungs

Natural human lungs are discussed in Sect. 5.13.2. The very high surface/volume ratio of the lungs (due to the presence of hundreds of millions of tiny alveoli) facilitates efficient gas exchange and permits respiration. Le Chatelier's Principle (named after French chemist Henry Louis Le Chatelier) is an equilibrium law; it predicts that if the concentration of a particular chemical species is high in one region and low in a connected region, the chemical will flow from the first region to the second to even out the concentrations. Because capillaries in the alveoli in the lungs have low concentrations of molecular oxygen (O_2), the high concentration of O_2 in air (20.95 mol%) allows oxygen to diffuse readily into the alveoli. In contrast, the concentration of carbon dioxide (CO_2) in the blood is higher than that in air (0.04 mol%), allowing CO_2 to diffuse into the alveolar space and be expelled from the body.

Remarkable as they are, lungs are subject to failure. Pal (2014) listed a large number of common lung diseases and conditions: chronic obstructive pulmonary disease (COPD), emphysema, pneumonia, asthma, acute bronchitis, pulmonary fibrosis, pleurisy, lung cancer, tuberculosis (TB), acute respiratory distress syndrome (ARDS), hypersensitivity pneumonitis, pulmonary hypertension, pulmonary embolism, and severe acute respiratory syndrome (SARS). In many cases, these disorders can be treated by medication, supplemental oxygen, radiation, or minor surgery.

But if these treatments don't work, lung transplants may be necessary. Surgeons can perform lobe, single-lung, double-lung, and heart–lung transplants. According to the *Scientific Registry of Transplant Recipients* issued by the U.S. Department of Health & Human Services, in 2017 there were 2478 lung transplants in the United States; 1360 people remained on the waiting list at the end of the year and 326 people (about 8%) died or became too sick for surgery before receiving a transplant. The report emphasized that, although there has been an increase in the number of donors in recent years, "the need for organs continues to outpace available donors."

There is thus a demand for artificial lungs. Heart–lung machines (a.k.a. cardiopulmonary bypass machines) are large external devices used during open-heart surgery when the heart is stopped. The machine intercepts blood before it enters the heart, pumps it to a reservoir where oxygen is infused, and circulates the oxygen-rich blood to the body's vital organs. After surgery, the machine is disconnected. It is not portable.

Small, portable, artificial lungs are used in non-surgical situations. These machines are designed to mimic lung function—adding O_2 to and removing CO_2 from the blood. But existing devices do an inadequate job; they achieve gas exchange rates far lower than natural lungs (0.25–0.40 L per minute vs. 2–6 L per minute) (Pal, 2014), leaving patients unable to resume a normal life. The principal reasons for this inefficiency are the lower surface/volume ratio and thicker membrane thickness of the artificial lungs.

There have been recent advances in the development of artificial lungs with higher surface/volume ratios. Thompson et al. (2017) described a small-scale "microfluidic" artificial lung produced by rolling a layer of siloxane (an organic silicone polymer) around a cylindrical membrane. [In microfluidic devices, fluids are maneuvered through geometrically small-scale channels where surface forces exceed volumetric forces.] The device described by Thompson et al. was able to achieve a high gas-exchange efficiency. The authors were optimistic that it could be scaled up to sizes suitable for human use.

Thompson et al. (2017) listed the performance characteristics of six microfluidic artificial lungs with different surface-area/blood-volume ratios of the effective gas-exchange area. There are two sets: those with low surface/volume ratios (≤ 125 cm^{-1}) had low CO_2 exchange rates (<200 ml min^{-1} m^2); those with high surface/volume ratios (≥ 800 cm^{-1}) had high CO_2 exchange rates ($\gtrsim 300$ ml min^{-1} m^2). The latter set is closer to the performance of natural lungs. Manufacturers are focusing on these devices.

9.3   Aerogel and *Stardust*

There are many synonyms for "extremely": e.g., enormously, exceedingly, exceptionally, extraordinarily, immensely, remarkably, very. All can be applied to aerogels.

Aerogels are rigid solid substances that are extremely porous (up to 99.98%); they have exceedingly low densities (as low as 0.000160 g cm^{-3}—more than seven times lower than the density of air at room temperature). As the name implies, aerogels are derived from gels, but instead of a liquid medium permeating a solid framework of tiny particles as in normal gel, the liquid has been replaced by gas. The solid particles in aerogel are typically 2–5 nm in diameter (where 1 nm = 10^{-9} m), fused together into clusters, forming a dendritic network. Pores are generally less than 20 nm across; the smallest pores are less than 1 nm.

Aerogels can be manufactured from diverse materials including silica (SiO_2), alumina (Al_2O_3), titania (TiO_2), zirconia (ZrO_2), chromia (Cr_2O_3), vanadia (V_2O_5), oxides of rare earth elements (neodymium (Nd), samarium (Sm), holmium (Ho), erbium (Er)), organic polymers, chalcogens (e.g., sulfur (S) and selenium (Se)), and graphene oxide (a compound containing carbon, oxygen, and hydrogen).

The most common type is silica aerogel, made from tiny particles of silica. With a porosity of 97% and a density as low as ~0.001 g cm^{-3}, silica aerogel has a remarkably low thermal conductivity (it is a great thermal insulator). It also has the slowest speed of sound of any solid substance. It is immensely strong relative to its mass—an aerogel block can support an object 2000 times heavier than the aerogel itself. Silica aerogel has a slightly blue color due to Rayleigh scattering; photons of short wavelength (i.e., blue light; 450–495 nm) are scattered in the aerogel by the dendritic network of small silica particles.

Because of its very high porosity, a chunk of aerogel would have an enormous surface area and a very high surface/volume ratio (orders of magnitude greater than that of a smooth, non-porous cube of the same size).

Other extreme properties of aerogel include an exceptionally low optical index of refraction (a characteristic derived from the ratio of the speed of light in a vacuum to that in a material) as well as an extraordinarily low dielectric constant (a measure of the ease of charge separation, and consequent change in voltage, of an insulating material relative to that in a vacuum).

The extremely high surface/volume ratio of silica aerogel was put to good use in NASA's *Stardust* mission, launched in February 1999 (Fig. 9.3). The principal focus of the mission was to collect particles from a cometary coma (the cloud of gas and dust expelled from a warming, devolatilizing comet) and return them to Earth. Ninety blocks of aerogel were mounted in a tennis-racket-shaped collector tray (Fig. 9.4), providing a large surface (>1000 cm^2) to entrap coma particles and interstellar dust grains. On 2 January 2004, the 390-kg spacecraft flew by Comet Wild 2 (pronounced "Vilt Two") at a distance of 237 km and a relative velocity of 6.1 km s^{-1}.

Cometary grains impacted the aerogel at initial velocities of ~6 km s^{-1}, interstellar grains collided at velocities up to ~30 km s^{-1}. The grains decelerated as they ploughed through the aerogel, forming long narrow conical tracks shaped like carrots. Many tracks were about 200 times longer than the particle's diameter. The extended release of kinetic energy during the collision with the aerogel allowed many of the impacting particles to remain intact. The grains, ranging from submicrometer size up to ~1 mm, came to rest at the end of the tracks. Perhaps a million cometary particles were captured as well as a few dozen interstellar grains.

Fig. 9.3 An artist's rendering of the Stardust spacecraft in the coma of Comet Wild 2. NASA image

Fig. 9.4 Stardust collector tray with aerogel-packed cells. NASA image

The cometary particles trapped by the aerogel are quite diverse. Some formed at high temperatures: grains of olivine $(Mg,Fe)_2SiO_4$, anorthite $(CaAl_2Si_2O_8)$, and diopside $(CaMgSi_2O_6)$; small aluminum-rich chondrules, FeO-rich chondrules and chondrule fragments; small calcium-aluminum-rich inclusions and fragments (CAIs; refractory-rich clasts containing Ca-, Al-, and Ti-rich minerals; Fig. 9.5); titanium nitride (TiN) and vanadium nitride (VN); and nuggets containing molybdenum (Mo) and several platinum-group elements (osmium (Os), iridium (Ir), ruthenium (Ru), platinum (Pt), rhodium (Rh)). Some captured materials formed at lower temperatures: organic compounds (including the amino acid glycine, $C_2H_5NO_2$; methylamine, CH_3NH_2; ethylamine, $CH_3CH_2NH_2$); and sulfides rich in iron, copper, and zinc (that likely formed in the presence of liquid water).

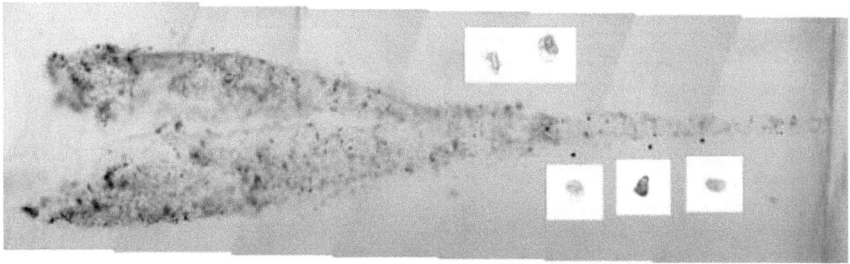

Fig. 9.5 The track in aerogel from the *Stardust* mission, produced by an impacting CAI particle. NASA image

The high-temperature materials were formed in the inner Solar System and must have been transported to great distances early in Solar System history—out to the zone where icy planetesimals (comets) were accreting. At some point, these planetesimals were heated sufficiently to melt some ice; the liquid water interacted with high-temperature grains to form iron- and copper-sulfide.

Down-to-Earth uses of aerogel include thermal insulation for retrofitting historic buildings and in manufacturing cycling gloves, thickening agents in paints, flexible materials for blankets and tennis rackets, and the construction of supercapacitors (found in laptops, defibrillators, and airplane evacuation slides). A market research report projects that the global aerogel product market will increase from 620 million dollars in 2019 to two billion dollars by 2023.

9.4 Giant Telescope Mirrors

Hans Lipperhey (c. 1570–1619) was a Dutch spectacle maker, widely credited with inventing the telescope. He filed a patent for the refracting telescope in 1608 but did not receive it because of competing claims of its invention. Lipperhey described his instrument as having been designed "for seeing things far away as if they were nearby". This basic telescope is a tube within which there is a convex objective lens that bends incoming parallel rays of light and brings them to a focal point. From there the light rays spread out a bit before entering a second lens (the eyepiece) where they are bent anew and become parallel again. The eye at the end of the tube is presented with a brighter, more-magnified image. Galileo heard about this invention in 1609 and built his own astronomical telescope to scan the heavens (Fig. 9.6). He observed mountains and craters on the Moon, noted the phases of Venus, discovered the four largest moons of Jupiter, found "ears" or "handles" on Saturn (deduced by Christian Huygens in 1655 to be rings), viewed sunspots (previously known only from naked-eye observations), and found the Milky Way to be awash in stars.

A major problem with early refracting telescopes was they suffered from chromatic aberration, essentially the failure of the objective lens to focus all wavelengths

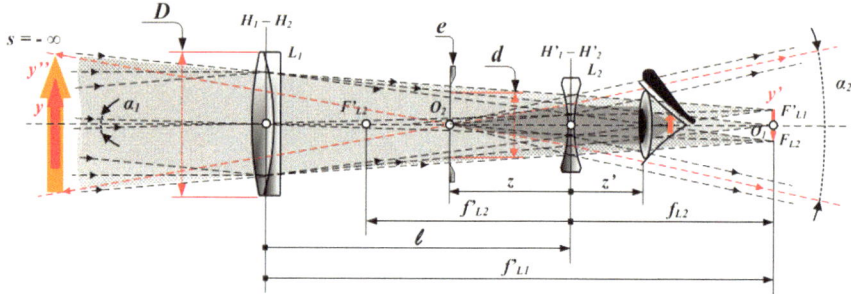

Fig. 9.6 Schematic diagram of the optical component of a Galilean telescope. It consists of an objective lens L_1 (either bi-convex or plano-convex) and an eyepiece lens L_2 (in this case bi-concave). Other symbols: y = extended view of a distant object; y' = image of the object from the objective lens; y'' = magnified virtual image from the eyepiece. *Image* by Tamasflex

of light to the same point. The image was blurry and exhibited colored rainbows flanking regions of high contrast.

Over the years, the optics improved, and portable refractors became commercially available (Fig. 9.7). Meanwhile, refracting telescopes kept getting larger. They reached their practical limit in 1897 with the 40-in. (102-cm) refractor at Yerkes Observatory in southern Wisconsin. At the time, it was the world's largest telescope.

But back in the Mid-Seventeenth Century, Isaac Newton was dissatisfied with the refracting telescopes then in use and took up the challenge in 1668 to construct the first reflecting telescope (Fig. 9.8). Parallel light rays entered a long tube and struck a curved, highly polished primary mirror at the bottom. The mirror, ground into the shape of a spherical cap, focused the light back up the tube. Before reaching the focal point, the light rays encountered a small, secondary mirror that reflected the concentrated light and focused it on the eye.

Because reflecting telescopes can accommodate very large mirrors, almost all major optical research telescopes in use today are reflectors. In nearly every modern

Fig. 9.7 Reverse of a 1-dollar commemorative coin from Australia depicting two young people peering at the night sky through a refracting telescope. The coin was issued for the International Year of Astronomy 2009. Also shown is the Southern Cross. Silver proof coin issued by the Perth Mint

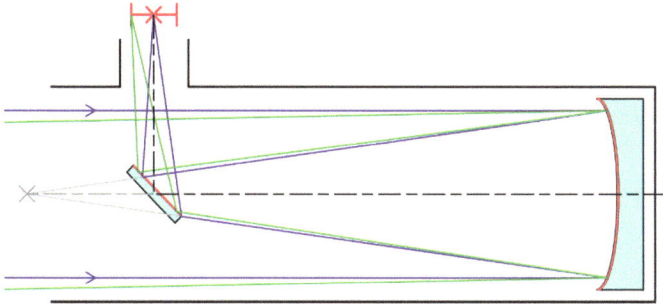

Fig. 9.8 Diagram of the optical path in a Newtonian telescope. The primary mirror at the bottom of the tube focuses the light back up the tube where it is intercepted before the focal point by a small secondary mirror that focuses the concentrated light out the side of the tube to the eye. *Image* by Krishnavedala

telescope, the primary mirror is a solid glass disk ground into the shape of a spherical cap or a parabola. The mirror is typically covered with a thin layer of aluminum.

Just as a large bucket will collect more water than a small bucket during a rainstorm, a large mirror will collect more light than a small mirror on a clear, moonless night. In optical astronomy, it is all about how much light can be collected—larger mirrors collect more light and provide greater sensitivity and higher resolution; they allow very faint objects to become observable.

Mount Wilson, rising 5713 feet (1741 m) in Southern California's Sierra Madre Range, is the site of the famous 60-in. reflecting telescope. The primary mirror started off as a 1900-pound (860-kg), 60-in. diameter (1.52-m) disk of plate glass, 7.5 in. (19-cm) thick, sitting in the basement of Yerkes Observatory in Wisconsin. The glass disk, modeled as a right circular cylinder, has a volume of 345,000 cm^3, a total surface area of 45,000 cm^2, and a surface/volume ratio of 0.13 cm^{-1}. After the mirror's arduous trip to California and the construction of Mount Wilson Solar Observatory, the telescope was assembled. The completed telescope was so heavy it required 20 metric tons of moving parts to keep the system operating smoothly during long exposures. The 60-in. telescope saw first light in December 1908, briefly becoming the largest telescope in the world.

But plans were afoot to produce a much larger reflecting telescope on Mount Wilson, one with a massive 100-in. (2.5-m) primary mirror. The glass disk weighed 9000 pounds (4100 kg); it was 101 in. (256 cm) across and 12 in. (30 cm) thick at the edge. The volume of the glass disk was 1.569 million cm^3, its total surface area was 127,000 cm^2, and its surface/volume ratio was 0.08 cm^{-1}, appreciably smaller than that of the 60-in. disk. However, the 100-in. glass disk was flawed—there were bubbles between different glass layers, formed when the layers of molten glass were poured into the mold. It took five years of grinding and polishing, but in 1916, the mirror was finally finished. The telescope saw first light in November 1917.

Among the many discoveries made with this remarkable telescope, two stand out—both made by Edwin Hubble.

First some background: Luminous clouds in space called "spiral nebulae" had been observed for years, but their nature was unknown. Were these relatively nearby clouds of gas and dust or were they "island universes"? If the latter, it would mean that the Milky Way—our galaxy—was just one of many and the universe was far larger than had been previously imagined. Not all stars shine at a constant rate; some are variable in brightness. One particular type is called a Cepheid variable. This type of star varies periodically in diameter and temperature. In 1908 Henrietta Swan Leavitt of Harvard College Observatory discovered a strong correlation between the pulsation periods of Cepheid variables and their intrinsic luminosity. Once the pulsation period was established, the star's intrinsic luminosity could be determined. The apparent luminosity of the star was a function of its distance (if potential dimming by interstellar dust is neglected).

Hubble used the 100-in. telescope to examine Cepheid variables in spiral nebulae and found the nebulae were too far away to be part of the Milky Way. The announcement was made in 1925 at the meeting of the American Astronomical Society. There were many galaxies in space. The universe was far larger than our home galaxy.

More background: Individual elements have characteristic emission lines at specific wavelengths due to their electrons moving from high-energy to lower-energy levels and emitting photons with these particular wavelengths. If there is a cool gas between the light source and the observer, these specific photons are absorbed by the gas, leaving dark lines (absorption lines) against the bright background of the visible spectrum. Objects moving toward or away from an observer have the wavelengths of their light shifted, either toward the blue end of the spectrum for approaching objects or toward the red end for receding ones. This is called a Doppler shift. If a luminous body is moving away, it takes longer for subsequently emitted waves of light to reach an observer; if the body is moving closer, the light waves are compressed, and it takes less time for subsequently emitted waves to reach the observer.

Hubble continued his observations with the 100-in. telescope, measuring distances to different galaxies. He found that the farther a galaxy was from the Milky Way, the faster it was moving away. He announced his findings in 1929: the universe was expanding. This is the principal basis for the Big Bang model of the origin of the universe. Yesterday, the galaxies were closer together than they are today. At some point in the past (now estimated at about 13.8 billion years ago), all the matter and energy in the universe was crammed into a single point.

The 200-in. (5.1-m) mirror at the heart of the telescope at Palomar Observatory in Southern California was cast in borosilicate glass (Pyrex). This material expands or shrinks much less than ordinary soda-lime-silica plate glass when experiencing temperature variations. The 200-in. mirror is attached to a honeycomb mount; this structure provides so much support that the amount of glass could be cut in half—from 36 to 18 metric tons.

Although the 100-in. and 200-in. telescopes facilitated great advances in astronomy, heavier glass disks tend to sag slightly under their own weight. This is because large surface areas (ideal for collecting more light) increase as the square of length, but volume and mass both increase as the cube. The sagging of the glass changes the mirror shape which must not vary by more than 50 nm to maintain its high resolution.

Today, larger glass mirrors are composites made from several smaller mirrors working in tandem. The W. M. Keck Observatory at the summit of Mauna Kea in Hawaii has twin telescopes, each with a 10-m-diameter mirror composed of 36 hexagonal segments. The positions of the segments are continuously adjusted by computer. Each segment is 1.8 m in diameter and 7.5 cm thick, and weighs about 450 kg. The volume of each segment is 191,000 cm^3, its surface area is 55,000 cm^2, and its surface/volume ratio is 0.29 cm^{-1}. Because mass and volume vary together as three-dimensional quantities, these smaller segmented mirrors provide more surface area, and hence, greater light-gathering capacity, for less weight than huge individual primary mirrors. The surface/volume ratio requires all giant optical telescopes using glass mirrors to be made of multiple mirrors (Fig. 9.9).

Another way to avoid the surface/volume problem of giant optical telescopes is to avoid glass entirely. The International Liquid-Mirror Telescope (ILMT) is a new four-meter-diameter telescope in India that saw first light in May 2022. It sits within an observatory building with a retractable rectangular roof. The ILMT uses a thin film of liquid mercury as its reflecting surface. The mercury sits atop a base that rotates once every eight seconds, shaping the liquid into a smooth parabolic shape. A thin transparent film of optical-quality mylar protects the mercury from wind. Light reflected from the mirror surface goes through a multi-lens optical corrector to produce sharp images. A large electronic camera sits at the focus. The telescope is stationary, pointing toward the zenith; it can observe only those objects that pass directly overhead. Nevertheless, potential observational targets abound, including asteroids, galaxies, gravitational lenses, and supernovae. If the telescope proves successful, many more will be constructed—they are much cheaper to build than giant glass mirror telescopes.

9.5 Nanoparticles

Onward and downward.

A common definition of a nanoparticle is one with dimensions between 1 and 100 nm. 1 nm = 10^{-9} m = 10^{-7} cm = 10^{-6} mm = 10^{-3} μm = 10 Å. For comparison: a strand of human DNA is ~ 2 nm in diameter; a gold atom is 0.288 nm in diameter; and common bacteria are 1000–2000 nm in diameter and 5000–10,000 nm long. A fingernail grows at a rate of about 1 nm per second. The lower limit of a nanoparticle would be the diameter of the hydrogen atom (roughly 0.1 nm—the width of the spherical electron cloud surrounding the proton in the nucleus). As pointed out by the U.S. National Nanotechnology Initiative, if the diameter of a marble were 1 nm, the diameter of the Earth would be ~1 m.

Nanometer-size particles have very high surface/volume ratios (Table 9.1). The surface/volume ratio of an 8-mm-diameter pea (750 m^{-1}) is 8 million times smaller than that of a spherical particle 1 nm across.

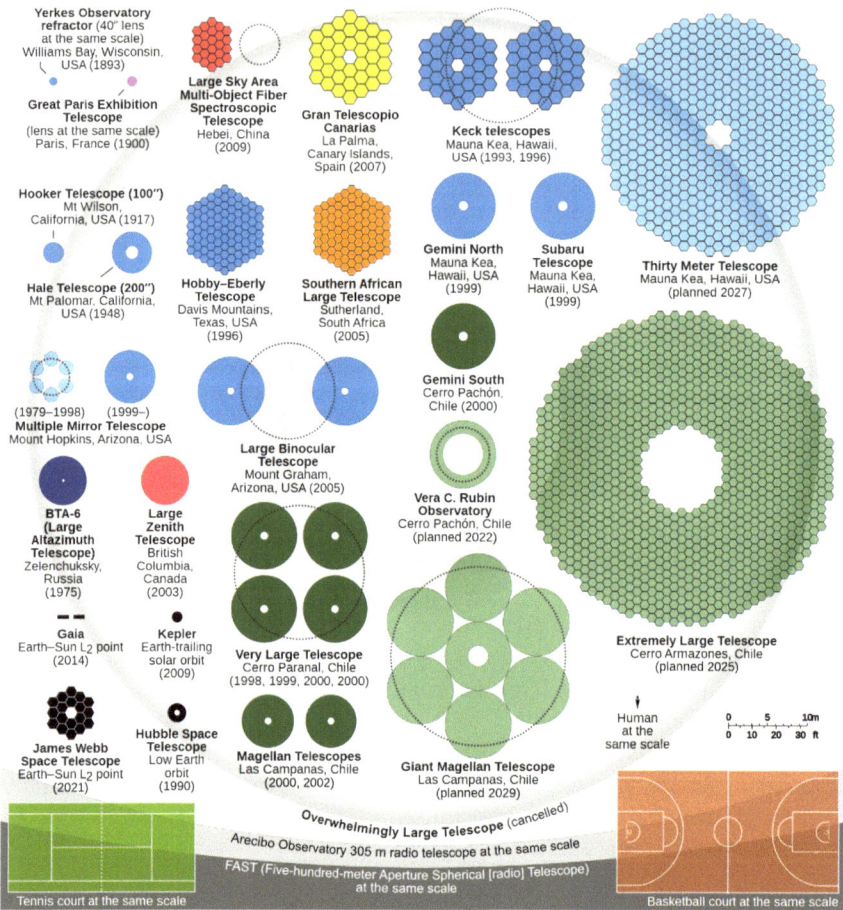

Fig. 9.9 Primary mirrors of large optical reflecting telescopes. The larger telescopes all use multiple mirrors working in tandem. *Image* by Cmglee

Because of their extremely high surface/volume ratios, nanoparticles are nearly all surface enclosing very little volume. At these small sizes, surface forces dominate three-dimensional bulk forces. Nanoparticles thus exhibit different physical properties than larger objects. For example, at very small size ranges, there is a systematic decrease in melting temperature. Nanometer-size gold particles melt at a temperature hundreds of degrees lower than that of bulk gold (Fig. 9.10).

This phenomenon, known as melting-point depression, is readily accounted for. Surface atoms have fewer neighboring atoms, less cohesive energy, and more unsatisfied chemical bonds than internal atoms, rendering them more chemically active. Surface atoms thus have higher surface energy and increased vibrational instability. On highly curved surfaces such as spherical nanoparticles, the amount of bonding available to a surface atom is reduced relative to that of a cubic particle of similar

Table 9.1 Geometric properties of spherical particles of different sizes

Particle diameter (nm)	Particle diameter (alternative unit)	Volume (nm^3)	Surface area (nm^2)	Surface/volume (nm^{-1})	Surface/volume (m^{-1})
1	0.001 μm	5.24×10^{-1}	3.14	5.99	5.99×10^9
10	0.01 μm	5.24×10^2	3.14×10^2	5.99×10^{-1}	5.99×10^8
10^2	0.1 μm	5.24×10^5	3.14×10^4	5.99×10^{-2}	5.99×10^7
10^3	1 μm	5.24×10^8	3.14×10^6	5.99×10^{-3}	5.99×10^6
10^6	1 mm	5.24×10^{17}	3.14×10^{12}	5.99×10^{-6}	5.99×10^3
10^7	1 cm	5.24×10^{20}	3.14×10^{14}	5.99×10^{-7}	5.99×10^2
10^9	1 m	5.24×10^{26}	3.14×10^{18}	5.99×10^{-9}	5.99
10^{12}	1 km	5.24×10^{35}	3.14×10^{24}	5.99×10^{-12}	5.99×10^{-3}

Fig. 9.10 The melting-point temperatures of small gold particles of different sizes. Very small particles melt at much lower temperatures than bulk gold (1338 K). *Diagram adapted* from Buffat and Borel (1976)

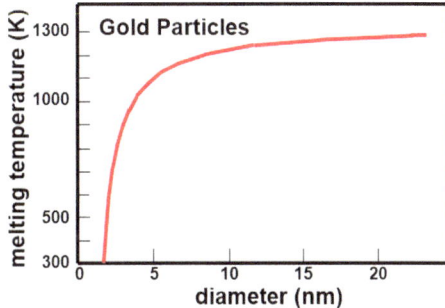

size. In a 3-nm-diameter spherical metal particle, approximately 90% of the atoms reside at the surface.

Melting occurs when sufficient heat is added to a solid allowing the thermal vibrations of atoms to overcome interatomic forces. Surface atoms detach from the solid. If a sample has an enormous surface/volume ratio and is composed almost entirely of surface atoms, less heat is required to cause the atoms to separate from the solid and enter the liquid state.

There is another property that changes with size: color. Collections of spherical silver metal particles change in color from cornsilk (120 nm) to azure sky blue (80 nm) to slate blue (40 nm). A collection of ~100-nm gold metal spheres is tangerine in color, but a collection of ~50-nm gold spheres is olive green. Cardell and Guerra (2022) examined gilded tin plasterwork in the Alhambra, an Islamic palace-fortress complex in Granada, Spain. They found the plasterwork had suffered corrosion and had spontaneously produced ~70-nm-diameter gold nanospheres that appeared purple.

Such color changes are a consequence of metallic bonding. Metal cations are compacted tightly in the crystal lattice; their valence electrons are not confined to clouds around a single nucleus but instead roam freely throughout the bulk metal. Such "delocalized" electrons move in response to applied electric fields. If these fields oscillate (e.g., like those of incident photons), the delocalized electrons also oscillate. These oscillations (called plasmons) occur at specific frequencies. Surface plasmons (confined to the surface of the metal) are functions of particle size, shape, and composition. Gold and silver have plasmon emissions in the visible spectrum, so when white light strikes gold and silver nanoparticles, the wavelengths corresponding to the specific plasmon frequencies of these metals are absorbed. The remaining, unabsorbed light frequencies are reflected, resulting in a change of color.

A potentially important use of nanotechnology is the separation of water molecules from impurities. This is accomplished naturally in cell membranes in bacteria, plants, and animals by aquaporin proteins, known as water channels. These channels facilitate the transport of water molecules across the membrane. Because of the narrowness of the channels, the surface/volume ratios of the openings are very high, and the rates of water flow are very low. Itoh et al. (2022) created fluorine-bearing nano-rings with interior diameters ranging from 0.9 to 1.9 nm. When inserted into phospholipid bilayer membranes, these rings formed nanochannels covered with fluorine atoms. The high electronegativity of the fluorine breaks water clusters apart, enabling individual water molecules to flow through the channels about 100 times faster than the rate of transport through aquaporins. When perfected, this same technology could be used to separate water from salt, creating the possibility of efficient desalination.

Nanotechnology was enabled by the development in 1981 of the scanning tunneling microscope (STM). The conducting, extremely thin tip of this microscope is brought near the surface of a target object. An applied voltage causes electrons to tunnel through the vacuum separating the tip and the target. Images can be produced as the tip scans the target surface, allowing features smaller than 0.1 nm to be distinguished. Individual atoms can be observed and manipulated by this device. However, a severe limitation is that the STM can make images only of conducting or semiconducting surfaces.

In 1985, a different instrument, the atomic force microscope (AFM), was developed to image any type of surface including glass, ceramic, and biological specimens. The AFM creates a three-dimensional image as it scans the surface of the target. The microscope's tip moves in response to electromagnetic forces between the tip and the atoms at the target surface.

Among the early products of nanotechnology are nanofibers—covalently bonded polymer chains made from various materials. Nanofibers have been used in tennis rackets, optical sensors, aircraft wings, air-filtration systems, tissue engineering, lithium-air batteries, and drug delivery systems.

But two types of nanoparticles have received the most attention: fullerenes and carbon nanotubes, both made solely from carbon atoms.

Fullerenes consist of rings of five to seven carbon atoms (connected by single and double bonds) forming a closed or partially closed cage. The most common type of fullerene is a hollow, closed cage, typified by buckminsterfullerene—a 0.71-nm-wide soccer-ball-shaped molecule made from 60 carbon atoms (C_{60}; Fig. 9.11). It is a truncated icosahedron, constructed from 20 hexagons and 12 pentagons. Its

Fig. 9.11 An illustration of a C_{60} molecule, a.k.a. a buckyball. *Image* by Michael Ströck

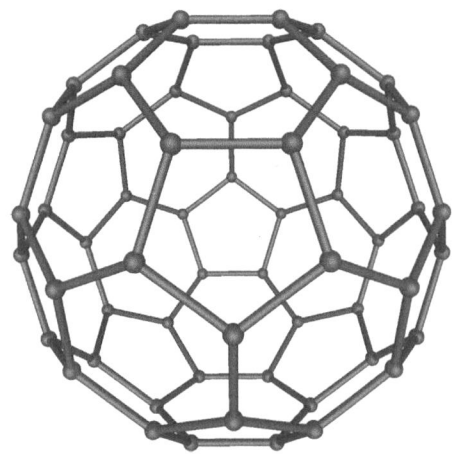

informal moniker is "buckyball". Both the formal and informal names honor American architect and designer Buckminster Fuller (1895–1983), inventor of the geodesic dome.

Buckminsterfullerene is the most common, naturally occurring fullerene. On Earth, it is found in soot and in black lustrous mineraloids called shungites; it is also produced by lightning in the atmosphere. In space, buckminsterfullerene has been observed in planetary nebulae and in a few unusual types of stars; ionized C_{60} was detected in the interstellar medium. Other fullerenes include those with 70, 72, 76, 78, 82, 84, and 100 carbon atoms. The smallest possible fullerene is C_{20}. One of the largest fullerenes synthesized so far is C_{3996}.

Ethylene glycol ($C_6H_6O_2$) is widely used in feedstock and in the manufacture of antifreeze and polyester fibers. It is synthesized on an industrial scale from dimethyl oxalate ($C_4H_6O_4$) over a palladium catalyst at pressures of about 20 bars. Zheng et al. (2022) showed the addition of C_{60} to a copper-silica catalyst allows ethylene glycol ($C_6H_6O_2$) to form from dimethyl oxalate at ambient pressure and moderate temperatures (180–190 °C). The reaction is highly efficient, with a yield of 98%.

Endohedral fullerenes are those with enclosed atoms, ions, or small molecules for use as superconductors or drug carriers. Nested cages of buckyballs ("buckyonions") have been proposed for use in lubricants. Disulfide analogs of fullerenes have been prepared with molybdenum (MoS_2), titanium (TiS_2), tungsten (WS_2), and niobium (NbS_2).

Carbon nanotubes are cylindrical fullerene molecules (sometimes called buckytubes) with a regular hexagonal lattice structure. The vertices are occupied by carbon atoms (Fig. 9.12). The nanotubes are typically a few nanometers across and range in length from less than a micrometer to several millimeters. They have high tensile strength and great flexibility; they exhibit high electrical conductivity (some forms are semiconductors), and high thermal conductivity. They are hydrophobic (they do not dissolve readily in water) and have been used in preparing stain-resistant

Fig. 9.12 Single-walled
carbon nanotube constructed
from bonded carbon atoms.
Image captured from an
animation in the German
Wikipedia

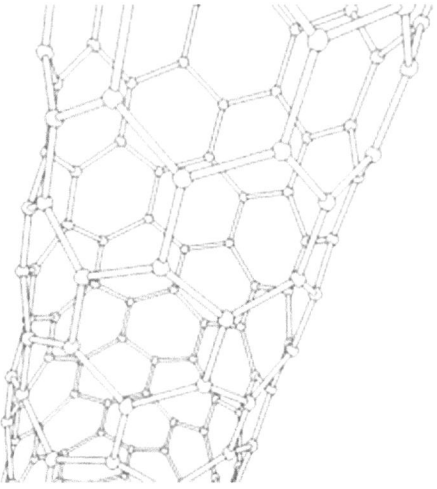

textiles. Carbon nanotubes have also proven valuable in the manufacture of optics and electronics.

Nanoproducts under development include carbon nanotube body armor for law-enforcement officers, bandages with interwoven silver nanoparticles to impede bacterial growth, and solar cells with a layer of porous titanium dioxide nanoparticles (a less-expensive alternative to crystalline silicon). Already on the market is Nanorepel™, a hydrophobic spray that uses a fine coating of quartz glass to protect glass and ceramic surfaces. It is advertised as "a user-friendly, ready-to-use formulation for coating … surfaces such as windows, mirrors, winter gardens, glass porches, glass tables, shower screens, wash basins, toilets, tiles, etc."

But nanoproducts pose potential health risks. One study of rats found that inhaled nanoparticles became lodged in their brains and lungs. The needle-like fibers of carbon nanotubes resemble those of asbestos and have been shown to produce lesions in the mesothelial lining of the chest cavity in mice (Poland et al., 2008). Other researchers showed that most of the silver nanoparticles that had been added to socks to reduce foot odor (by preventing bacteria from reproducing) were lost after only a few washings. The nanoparticles were released with the wastewater into the environment, potentially harming fish and ecologically critical bacteria.

In 2004, the Royal Society and the Royal Academy of Engineering issued a joint report, *Nanoscience and Nanotechnologies: Opportunities and Uncertainties*, highlighting additional health concerns. The report noted that "the relatively greater surface area of nanoparticles, given equal mass, and their probable ability to penetrate cells more easily and in a different way" is a potential health concern. It warned that "some manufactured nanoparticles and nanotubes are likely to be more toxic per unit mass than particles of the same chemicals at larger size and will therefore present a greater hazard…." The report concluded that nanoparticles could potentially "overcome the body's natural defenses."

Nevertheless, a December 2021 BCC Research report forecast that the global market for nanotechnology would grow from \$5.2 billion in 2021 to \$23.6 billion by 2026.

References

Buffat, Ph., & Borel, J.-P. (1976). Size effect on the melting temperature of gold particles. *Physical Review A, 13*, 2287–2298.

Cardell, C., & Guerra, I. (2022). Natural corrosion-induced gold nanoparticles yield purple color of Alhambra palaces decoration. *Science Advances, 8*, eabn2541.

Hoppe, A., Güldal, N. S., & Boccaccini, A. R. (2011). A review of the biological response to ionic dissolution products from bioactive glasses and glass-ceramics. *Biomaterials, 32*, 2757–2774.

Itoh, Y., Chen, S., Hirahara, R., Konda, T., Aoki, T., Ueda, T., Shimada, I., Cannon, J. J., Shao, C., Shiomi, J., Tabata, K. V., Noji, H., Sato, K., & Aida, T. (2022). Ultrafast water permeation through nanochannels with a densely fluorous interior surface. *Science, 376*, 738–743.

Li, B., & Aspden, R. M. (1997). Composition and mechanical properties of cancellous bone from the femoral head of patients with osteoporosis or osteoarthritis. *Journal of Bone and Mineral Research, 12*, 641–651.

Nguyen T. D., Kadri O. E., Sikavitsas V. I., & Voronov, R. S. (2019). Scaffolds with a high surface area-to-volume ratio and cultured under fast flow perfusion result in optimal O_2 delivery to the cells in artificial bone tissues. *Applied Science, 9*, 2381. https://doi.org/10.3390/app9112381

Pal, S. (2014). *Design of artificial human joints & organs*. Springer, 419 pp.

Poland, C. A., Duffin, R., Kinloch, I., Maynard, A., Wallace, W. A. H., Seaton, A., Stone, V., Brown, S., MacNee, W., & Donaldson, K. (2008). Carbon nanotubes introduced into the abdominal cavity of mice show asbestos-like pathogenicity in a pilot study. *Nature Nanotechnology, 3*, 423–428.

Saijo, H., Fujihara, Y., Kanno, Y., Hoshi, K., Hikita, A., Chung, U., & Takato, T. (2016). Clinical experience of full custom-made artificial bones for the maxillofacial region. *Regenerative Therapy, 5*, 72–78.

Thompson, A. J., Marks, L. H., Goudie, M. J., Rojas-Pena, A., Handa, H., & Potkay, J. A. (2017). A small-scale, rolled-membrane microfluidic artificial lung designed towards future large area manufacturing. *Biomicrofluidics, 11*, 024113, 12. https://doi.org/10.1063/1.4979676

Zheng, J., Huang, L., Cui, C.-H., Chen, Z.-C., Liu, X.-F., Duan, X., Cao, X.-Y., Yang, T.-Z., Shi, H. Z., Du, P., Ying, S.-W., Zhu, C.-F., Yao, Y.-G., & Guo G.- C., Yuan Y., Xie S.-Y. and Zheng L.-S. (2022). Ambient-pressure synthesis of ethylene glycol catalyzed by C_{60}-buffered Cu/SiO_2. *Science, 376*, 288–292.

Epilogue

Philosophical physicalism (also known as philosophical materialism, ontological naturalism, and metaphysical naturalism) is the metaphysical proposition that everything in the universe is a physical entity. This includes spacetime, matter, energy, dark matter, and dark energy; it includes waves, electromagnetic fields, and quantum fields; it encompasses all interactions among these entities and any properties that emerge from such interactions. The physicalist proposition holds that any features of the universe yet to be discovered are also physical entities. The universe itself is a physical entity and, if the multiverse exists, it and every component of it are entirely physical.

Thought, imagination, mental states, emotions, reasoning, consciousness, abstract concepts like justice, beauty, morality, love, and ethics, and disciplines and tools like mathematics, philosophy, and history are all covered. So are art, literature, and music. These are all either emergent properties of the physical world, natural activities produced by physical entities, or descriptions of such processes and activities. Physicalism embodies all of reality. Information, numbers, the laws of logic, physical laws, and physical constants are included. There is no room in the physicalist view for the supernatural—no gods, no angels, no demons, no souls, no spirits. The universe is understood as having no underlying purpose. If physicalism is true, the Sixth Circle of Hell (Heresy) is an empty place. (The other circles are not so crowded either.)

Modern science is less dogmatic. When engaged in their work, scientists adhere to the principal of methodological naturalism. They seek naturalistic explanations for phenomena. If no such explanations are forthcoming, they admit ignorance and do not invoke supernatural causes.

A glance at the Table of Contents shows that a single geometric property—the surface/volume ratio—governs much of the behavior of the physical world.

The size range is huge. It spans more than 17 orders of magnitude from a single atom (0.1 nm—the diameter of hydrogen) to the Earth (mean diameter $= 12{,}742$ km $= 1.2742 \times 10^{16}$ nm).

A. E. Rubin, *Surface/Volume*, https://doi.org/10.1007/978-3-031-23749-2

The variety of objects is enormous. It covers planets and asteroids, basalts and sandstones, birds and butterflies, sponges and elm trees, pizza and cotton candy. It explains the shape of the human brain and the structure of the small intestine.

The range of physical processes is wide. It includes melting, evaporation, diffusion, and osmosis, the dispersal of disrupted meteoroids, and geological erosion and weathering. It accounts for exploding grain elevators, the bleaching of coral reefs, and the spread of wildfires. It explains why tiny metal particles melt at relatively low temperatures and why different-sized grains of gold display different colors.

The shape and size of manufactured objects is also constrained by the surface/volume ratio. It shows why most giant reflecting telescopes have multiple mirrors, why aerogel was used to capture cometary dust, and why we need be wary of applying silver nanoparticles to our socks.

The universe is vast. Geometry encompasses it all. In Book VII of *The Republic*, Plato wrote that "The knowledge at which geometry aims is knowledge of the eternal. Geometry will draw the soul towards truth, and create the spirit of philosophy.... Nothing will be more likely to have such an effect. Nothing should be more sternly laid down than that the inhabitants of your fair city should by all means learn geometry." Johannes Kepler understood this intuitively (Fig. E.1). He wrote "Geometry is the archetype of the beauty of the world."

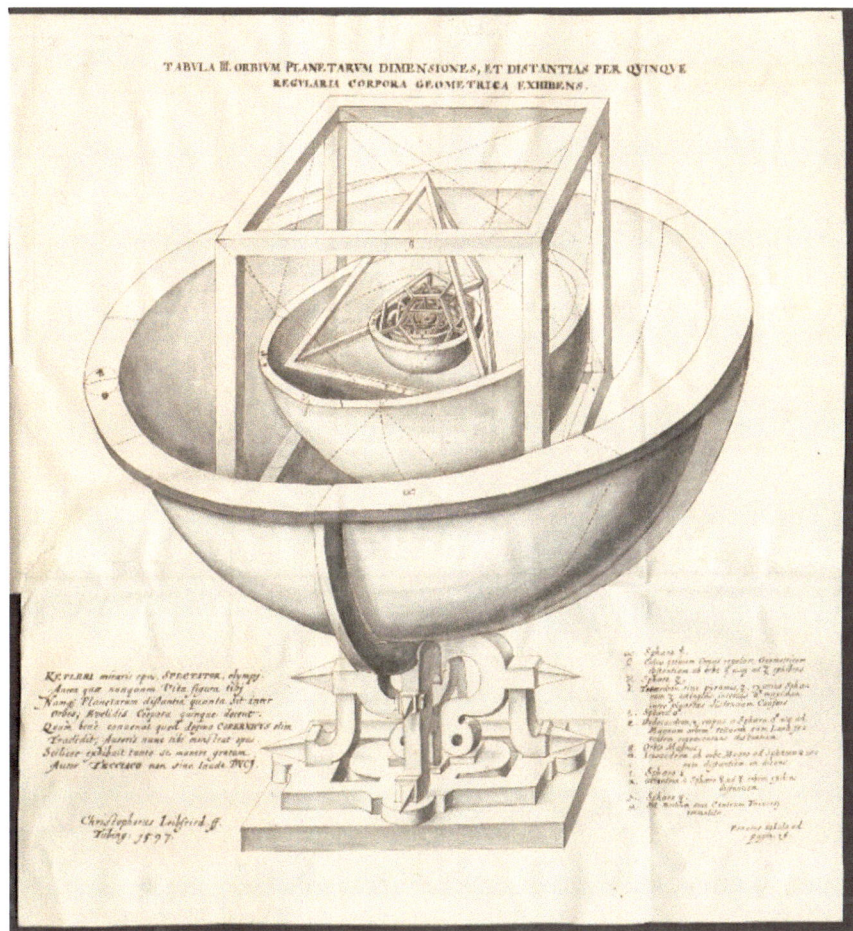

Fig. E.1 Platonic solids containing the orbits of the planets. Illustration from *Mysterium Cosmographicum* by Johannes Kepler (1597)

Index

A

Abdomen, 75, 83, 87
Abscission layer, 95
Aerogel, 158–161, 174
Albatross, 73
Algae, 151–153
Alimentary canal, *see* gastrointestinal tract
Allen's Rule, 79
Alveolar ducts, 101, 102
Alveoli, 101, 102, 157
Amino acids, 59, 95, 102, 103, 108, 117,
    119, 120, 151, 155
Anions, 126, 127, 132
Ants, 68, 75, 76
Anus, 87, 102, 104, 105
Aqueous alteration, 40, 41
Archimedes, 7, 63
Aristotle, 5, 7
Artificial bone, 156
Artificial lungs, 158
Asteroids, 22, 25–28, 30–33, 36, 39–41, 46,
    50, 165, 174
Astronomical unit, 25
Atomic force microscope, 168
Atomic nucleus, 125, 126, 128
Atomic number, 125, 126, 129
Atomic weight, 125, 129
ATP, 60, 61
Autumnal equinox, 95
Axons, 97

B

Bacteria, 61, 108–112, 165, 168, 170
Basalt, 31, 46–50, 174
Batholith, 46
Bats, 71, 74

Bergmann's Rule, 77–79, 81
Bile, 103, 104
Biofilm, 111
Birds, ix, x, 61, 65, 67–74, 77, 79, 83, 94,
    147, 174
Blood pressure, 133
Body mass, 62–66, 68–71, 76
Bolus, 102
Bone, x, 82, 155–157
Brain, xi, 97–99, 122, 174
Brainstem, 97, 98
Brine, 53
Bronchi, 99
Bronchioles, 99, 101, 102
Buckminsterfullerene, 168, 169
Bumblebees, 83
Butterflies, 174

C

Calcium-Aluminum-rich Inclusions
    (CAIs), 40, 41, 160
Capillary action, 53, 94
Carbohydrates, 59, 94, 151, 156
Carbonaceous chondrites, 40, 41
Carbon dioxide, 54, 60, 61, 76, 88, 93, 95,
    98, 101, 132, 138, 147, 152, 157
Carbonic acid, 54
Carbon nanotubes, 168, 170
Carotenoids, 95
Cartesian coordinates, 8
Cations, 126, 127, 132, 167
Cats, 61, 68, 141
Cecum, 103, 104
Cellulose, 59, 138, 143
Cephalothorax, 87
Cepheid variable, 164

Cerebellum, 97, 98
Cerebral cortex, 97
Cerebrum, 97, 98
Ceres, 27
Chalk, 9
Chelyabinsk meteorite, 33–36
Chemical bonding, 127
Chimpanzees, 68
Chitin, 75, 87
Chlorine, 132, 133
Chlorophyll, 93, 95
Chloroplasts, x, 93, 94
Chondrites, 36, 37, 40, 41
Chondrules, 36–42, 160
Chord, 15
Chyme, 102, 103
Circle, 1–3, 6, 8, 15, 17, 18, 36
Circumference, 5, 15–17
Coal, 45, 142
Colon, 104, 108
Comets, 25, 28, 161
Comet Wild 2, 159, 160
Common notions, *see* Euclid
Compact bone, 82, 155
Cone, 11
Congruent, 2, 4
Conifers, 94
Coral bleaching, 152, 153
Coral reefs, 151, 174
Corpus callosum, 98
Cosmic dust, 42
Cosmic spherules, 42, 43
Cotton candy, 136, 137, 174
Cougars, 78
Covalent bond, 127, 129, 131
Cows, 62, 68
Cube, x, 4, 12, 14, 15, 19, 22, 23, 63, 69,
        104, 140, 159
Cubit, 1
Cutin, 92
Cylinder, 7, 9, 62, 104, 148, 163

D
Deciduous trees, x, 23, 95, 96
Dehydration, 84, 93
Delocalized electrons, 168
Deoxyribonucleic Acid (DNA), 59, 151,
        165
Deposition, 52
Descartes, René, 7, 8
Diagenesis, 45
Diffusion, 39–41, 61, 101, 112, 113, 139,
        174

Dinoflagellates, 151
Dinosaurs, x, 82
Displacement, 62, 63
Dissolution, x, 131
Dogs, 10, 61, 64, 113
Doppler shift, 164
Drag, 35, 36, 72–74, 93, 95
Duodenum, 103
Dwarf planet, 27

E
Earth, ix, 2, 5, 10, 11, 17, 23, 25, 27–33, 42,
        43, 45, 46, 49, 52–55, 75, 82, 94,
        108, 132, 133, 159, 165, 169, 173
Ectotherms, 80
Einstein, Albert, 7
Electronegativity, 128, 129, 168
Electron orbitals, 127
Electrons, 125–131, 164, 167, 168
Electron shells, 126–128
Electrostatic attraction, 127, 128, 130–133
Elephants, x, 64–68, 82
Endotherms, 80
Epidermis, 92, 93
Epiglottis, 102
Epithelium, 101
Eratosthenes, 5
Erosion, 52, 174
Erythrocytes, 112
Esophagus, 102
Ethanol, 61
Euclid, 5, 6, 8
Eucrites, 30, 31
Evaporation, x, 62, 84, 94, 133, 136–138,
        174
Exoskeleton, 75, 76, 87, 88

F
Falcons, 68, 73, 74
Fast gait, 71–74
Feathers, 61, 67, 69, 79
Feces, 104
Feeding basket, 87, 88
Feldspar, 52, 54
Foraminifera, 45
Fracking, 50
Frogs, 86, 147
Frontal lobe, 98
Frostbite, xi, 106
Frost wedging, 53
Fructose, 132
Fudge, 135

Fuel availability, 148, 149
Fuel load, 149
Fullerenes, 168, 169
Fusion crust, 33

G

Gabbro, 48, 49
Galaxies, 11, 164, 165
Galena, 12
Galileo Galilei, 8–10, 161
Garlic, 19, 21
Gaster, 75
Gastrointestinal tract, 102
Gastrolith, 53, 71
Geometry, ix, x, 1, 5–8, 23, 122
Gerbils, xi, 68
Gigantothermy, 80
Glucose, 60, 61, 75, 132, 151, 155
Goose bumps, 105
Gorillas, 81
Grain elevator, xi, 141, 143, 174
Granite, 46, 50, 53, 54
Gulls, 73, 79

H

Halite, 12, 127, 133, 134
Hawks, 74
Heart-lung machine, 158
Heat of fusion, 131
Heat of vaporization, 62, 131, 137
Hemoglobin, 112, 113, 118, 120
Hesse's Rule, 79
HMS Challenger, 43
Hovering, 72, 74, 75
Hubble, Edwin, 163
Hummingbirds, x, 68, 69
Hydrogen bond, 131
Hydrophilicity, 117
Hydrophobicity, 117, 120, 121
Hydrostatic equilibrium, 25
Hydrothermal vents, 108
Hypertension, 134, 157

I

Ice segregation, 53, 55
Ichthyosaurs, 80
Igneous rocks, 48, 54
Ileum, 103, 104
Immiscible liquids, 27
Index of refraction, 159
Insects, 68, 71, 74–76, 82, 83

Insulin, 118, 119
Integument, 84
Intestinal villi, 103
Ionic bond, 127
Irrational number, 15
Isotopes, 30, 41, 43, 125

J

Jejunum, 103
Jupiter, 25, 26, 161

K

Keck Observatory, 165
Kepler, Johannes, 174, 175
Kīlauea, ix
King Kong, 81, 82
Kleiber's law, 65
Komatiite, 28
Krill, xi, 70, 87, 88
Kuiper Belt, 25, 28

L

Large intestine, 102–104
Larynx, 98, 102
Lattice, 127
Lava, ix, 30, 31, 46, 49, 136, 137
Leaves, x, 23, 68, 70, 71, 76, 77, 91–96,
     147, 148, 152
Leavitt, Henrietta Swan, 164
Lift, 72–74
Limestone, 45
Lincoln, Abraham, 6
Lipids, 113, 151, 155
Lungs, 75, 98, 99, 101, 102, 122, 157, 158,
     170

M

Macaws, 70
Magma, ix, 46–50
Magnetite, 14, 40, 41, 43
Mammals, ix, 61, 65, 67–69, 71, 77, 82, 147
Mantle depth, 32
Mars, 25–29, 31, 46
Medial longitudinal fissure, 98
Mediterranean Sea, 133
Medulla, 98
Mercury, 25, 28–31, 46
Mesohyl, 88–90
Mesostasis, 37
Metabolic rate, x, 64–69, 71, 80, 81, 83
Metabolism, 60, 61, 95

Metallic bond, 127, 167
Metamorphic rocks, 46, 54
Meteorites, 14, 27, 30, 33, 36, 38–40
Meteoroids, 33, 35, 174
Mice, *see* mouse
Microchondrules, 37–40
Micrometeorites, 42, 43
Microscope, 88, 108, 157
Midbrain, 98
Mitochondria, 60
Moon, ix, 5, 23, 27–31, 46, 161
Mosasaurs, 80
Moths, 83, 84
Mount Wilson Observatory, 163
Mouse, x, 36, 64–68, 141
Mouse-to-elephant curve, 65, 67
Muscles, x, 62, 72–75, 82, 83, 104, 105
Muscle mass, x
Myelin, 97
Myoglobin, 119

N
Nanofibers, 168
Nanoparticles, x, 165, 166
Neptune, 25, 26, 28
Newton, Isaac, 162
Nucleic acids, 59, 95

O
Obsidian, 46
Occipital lobe, 98
Olivine, 37, 38, 40, 42, 47, 48, 54, 160
Oort Cloud, 25, 28
Orbitals, 126, 127, 129
Ordinary chondrites, 36, 40
Osmosis, 139, 140, 174
Osteocytes, 155
Osteons, 155
Ostriches, 70

P
Palisade layer, 93
Palomar Observatory, 164
Pancreas, 103, 119
Panting, 62
Parietal lobe, 98
Peat, 45, 147
Pele's hair, 136, 137
Pelicans, 70
Penguins, 70, 77, 106
Perfect squares, 4

Perimeter, 15
Periodic table, 126, 128, 129
Peristalsis, 102, 104
Permeability, 50–52
Petiole, 75, 91, 95
Pharynx, 98, 102
Phloem, 94
Photosynthesis, x, 59, 60, 93–95, 151
Phyllosilicates, 40, 41
Physicalism, 173
Phytoplankton, 87
Pizza, 133, 134, 174
Planet, xi, 26, 28, 30, 32, 108
Plasmons, 168
Plate boundaries, 30
Plate tectonics, 31, 52
Plato, 174
Platonic solids, 175
Pluto, 28
Plutonic rocks, 46, 47
Polar bears, 79
Polygons, 2
Polypeptide, 59, 117–119
Pons, 98
Porosity, 50, 90, 156, 159
Postulates, *see* Euclid
Protein domain, 120
Proteins, 59, 88, 95, 102, 103, 113,
       117–120, 122, 138, 151, 155, 168
Pylorus, 102, 103
Pyramid, 2, 3, 134
Pyrite, 12
Pyrophytes, 149
Pyroxene, 37
Pythagoras, 1, 2, 4, 5
Pythagorean Theorem, 1, 4

Q
Quartz, 46, 47, 51, 52, 170

R
Radius, ix, 6, 8, 11, 15, 17–21, 32, 49, 90,
       120, 144
Rational number, 17
Rayleigh scattering, 159
Rectangular prism, 11
Rectum, 104
Red blood cells, 112, 113, 120, 155
Red oak, 91
Refractory inclusions, *see* CAIs
Respiration, 60, 62, 75, 84, 99, 157
Respiratory tract, 62, 98, 99, 101

Rhyolite, 46
Ribonucleic Acid (RNA), 59
Rock candy, 135
Running, 71

S

Sandstone, 45, 46, 50–52, 174
Saturation point, 133
Saturn, 26, 161
Sauropods, 82
Scanning electron microscope, 39
Scanning tunneling microscope, 168
Sedimentary rocks, 45, 50, 54
Sediments, 43, 45, 50, 52, 88
Seral forests, 149
Shale, 45, 46, 50
Shivering, ix, 83
Shrews, x, 67
Silica, 46, 54, 88, 159
Silo, 141, 142
Slow gait, 71–74
Small intestine, xi, 102–104, 122, 133, 174
Sodium, 14, 62, 132–134
Soil, 52, 56, 57, 108, 149
Soil profile, 56, 57
Solar nebula, 41
Solar System, ix, 25–32, 38, 161
Solvation, *see* dissolution
Specific heat capacity, 67
Sphere, 5, 7, 11, 17–21, 38, 90, 92, 98, 102,
        104, 118, 120, 126, 143, 144, 167
Spinal cord, 97, 98
Spiracles, 75
Sponges, 88–90
Spongin, 88
Spongy bone, 82, 155, 156
Spongy layer, 93
Square, 1, 4, 6, 12–14
Squid, 70, 151
Stadia, 5
Stardust mission, 159, 161
Stomach, 71, 102, 103
Stomata, 93, 95
Strait of Gibraltar, 133
Strewn fields, 33, 34
Sucrose, 132, 135, 142
Summer solstice, 5, 95
Sun, x, 5, 10, 25–28, 33, 38, 94
Surface area, ix, x, 1, 7, 8, 11, 12, 14–20,
        22, 23, 38, 49, 51, 55, 56, 62–64, 69,
        73, 79, 88–91, 94, 98, 101–106, 113,
        118, 120, 137, 138, 140, 143, 148,
        149, 156, 159, 163, 165, 167, 170

Surface/mass ratio, x, 63, 73, 79
Surface tension, 17, 38, 42, 131
Surface/volume ratio, x, 8, 16, 19, 20, 23,
        31, 32, 35, 38–40, 43, 49, 51, 52, 55,
        56, 63, 65, 67, 69, 71, 73, 79, 83–86,
        88–90, 92, 98, 102, 104, 106, 107,
        112, 113, 120, 122, 134, 136, 138,
        140, 148, 152, 155–159, 163, 165,
        167, 173, 174
Swans, 68, 70
Sweat glands, 62, 105
Sweating, 84

T

Telescope, x, 161–166, 174
Temporal lobe, 98
Thales, 2, 3
Thermal diffusivity, 50
Thermoregulation, 61, 67
Thorax, 75, 83, 87
Thrust, 72
Tigers, 78
Tortoises, 80
Tortuosity, 50–52
Trachea, 75, 76, 98, 99, 102
Transpiration, 94, 95
Triangle, 2, 3, 4, 6
Triangular prism, 11
Turbulence, 71
Turgor, 139, 140

U

Uranus, 26

V

Valence electron, 127
Vascular tissue, 94
Venus, 25, 28–30, 46, 161
Vertebrates, x, 63–65, 98, 112, 113
Vesta, 28–31, 46
Viscosity, ix, 28, 46, 50
Volcanic rocks, 28, 43, 46
Volcano, ix, 32, 49, 52
Volume, ix, x, 7, 8, 11, 12, 15–20, 22, 23,
        35, 36, 38, 49–51, 55, 56, 62–64, 69,
        84, 89–91, 98, 102, 104–106, 118,
        133, 137, 138, 148, 163–165, 167

W

Wackes, 52
Walking, 2, 71

Waste heat, 61, 83
Weathering, 53–57, 174
Wildfires, 147, 149, 174
Wing loading, 84
Winter solstice, 94

X
Xylem, 94, 131

Y
Yeast, 61
Yellowstone National Park, 32

Z
Zooplankton, 87